Lecture Notes
in Control and Information Sciences   214

Editor: M. Thoma

Springer
London
Berlin
Heidelberg
New York
Barcelona
Budapest
Hong Kong
Milan
Paris
Santa Clara
Singapore
Tokyo

George Yin and Qing Zhang (Eds)

Recent Advances in Control and Optimization of Manufacturing Systems

Springer

Series Advisory Board

A. Bensoussan · M.J. Grimble · P. Kokotovic · H. Kwakernaak
J.L. Massey · Y.Z. Tsypkin

Editors

George Yin, Dr
Department of Mathematics, Wayne State University
Detroit, Michigan 48202, USA

Qing, Zhang, Dr
Department of Mathematics, University of Georgia
Athens, Georgia 30602, USA

TS
155
.R3559
1996

ISBN 3-540-76055-5 Springer-Verlag Berlin Heidelberg New York

British Library Cataloguing in Publication Data
Recent advances in control and optimization of
  manufacturing systems. - (Lecture notes in control and
  information sciences ; 214)
  1.Production control 2.Production planning 3.Scheduling (Management)
  I.Yin, George II.Zhang, Qing, 1959-
  658.5
ISBN 3540760555

Library of Congress Cataloging-in-Publication Data
A catalog record for this book is available from the Library of Congress

Apart from any fair dealing for the purposes of research or private study, or criticism or review, as permitted under the Copyright, Designs and Patents Act 1988, this publication may only be reproduced, stored or transmitted, in any form or by any means, with the prior permission in writing of the publishers, or in the case of reprographic reproduction in accordance with the terms of licences issued by the Copyright Licensing Agency. Enquiries concerning reproduction outside those terms should be sent to the publishers.

© Springer-Verlag London Limited 1996
Printed in Great Britain

The publisher makes no representation, express or implied, with regard to the accuracy of the information contained in this book and cannot accept any legal responsibility or liability for any errors or omissions that may be made.

Typesetting: Camera ready by editors and authors
Printed and bound at the Athenæum Press Ltd, Gateshead
69/3830-543210 Printed on acid-free paper

PREFACE

In recent years, we have witnessed rapid progress in research of manufacturing systems. Many important results have been obtained. Very often these contributions are made through interdisciplinary efforts and collaborations. The theoretical findings, in turn, have significant impact on various applications. The applications, on the other hand, bring about new and exciting theoretical discoveries. In the very competitive world, research in manufacturing systems plays an important role in creating and improving technologies and management practices of the economy. The purpose of this volume is to present some most recent results in stochastic manufacturing systems. We have invited experts from engineering, management sciences, and mathematics. These authors review and substantially update the recent advances in control and optimization of manufacturing systems.

This volume consists of 8 chapters divided into three parts, which are optimal production planning, scheduling and improvability, and approximation of optimal policies. All papers have been refereed. To facilitate the reading, let us briefly go through the outline of the chapters below.

OPTIMAL PRODUCTION PLANNING. This part begins with the paper of Ching and Zhou. They consider optimal hedging policy for a failure prone system consisting of one-machine, which produces one type of product and with batch arrival demand. The machine states and the inventory levels are modeled as Markov processes. An algorithm is used to obtain the steady state probability distribution. The optimal strategy is then obtained by varying different values of hedging point.

The paper of Huang, Hu and Vakili deals with manufacturing systems in which the production capacity is allowed to be augmented by hiring temporary workers, adding overtime and/or extra shifts, or outsourcing part of the production. A production control problem is to strike a balance between the costs of building inventory, buying extra capacity, and backlogged demand. The optimal control policy is characterized by three switching levels.

In Chapter 3, Perkins and Srikant examine the optimal control of a fluid flow model for manufacturing systems. They manage to show that the optimal control can be obtained by solving a series of quadratic programming problems having linear inequality constraints. For the case in which the machines are failure-prone, they show that the steady-state probability densities of the buffer levels satisfy a set of coupled hyperbolic partial differential equations.

SCHEDULING AND IMPROVABILITY. The work of Maddox and Birge treats a scheduling problem for stochastic job shops. The authors propose a scheduling heuristic which performs restricted dynamic updating of an initial schedule for the shop, using limited distribution information about the random quantities. In the model, they allow general dependence among the stochastic quantities and require at most two moments of each distribution function. The objective is to minimize expected tardiness. They also investigate the benefits of the scheduling heuristic versus a dispatching rule using a simulation study.

Jacobs, Kuo, Lim and Meerkov investigate the improvability of scheduling manufacturing systems. They model the production systems with serial machining and assembly lines with unreliable machines and finite buffers. The limited resources involved are the total work-in-process (WIP) and the workforce (WF). The performance index addressed is the throughput. It is shown that the system is unimprovable with respect to the WF if and the only if each buffer is on the average half full. The system is unimprovable with respect to WIP and WF simultaneously if and only if the above holds and, in addition, each machine has equal probability of blockage and starvation. Thus, in a well designed system each buffer should be mostly half full and each machine should be starved and blocked with equal frequency.

APPROXIMATE OPTIMALITY AND ROBUSTNESS. It is well known that exact optimal control is often difficult to obtain. In order to control a real-time manufacturing system, one is often more interested in finding approximately optimal controls, and designing robust controllers. In Chapter 6, Sethi and Zhou discuss a stochastic two-machine flowshop. They propose a hierarchical control approach to construct a control for the flowshop that is asymptotically optimal with respect to the rates of changes in machine states. Comparisons between their methods and the frequently used Kanban controls are given.

Chapter 7 concentrates on a flow control model of a flexible manufacturing system. Zhu introduces a deterministic control problem to deduce equations governed by ordinary differential equations that determine the moments of the inventory levels. He discusses the methods to study the expected inventory level set and the optimal inventory level control problem with quadratic cost functions by using these control equations.

Chapter 8, written by Boukas, Zhang and Yin, is concerned with robust production and maintenance planning. By modeling the problem under minimax framework, they aim at designing controls that are robust enough to attenuate uncertain disturbances including modeling errors, and to achieve system stability.

In addition to the aforementioned topics, the authors describe the impact of the underlying study on many applications in manufacturing systems. They also suggest many new topics of research for future in-depth investigation.

Without the help of many people, this volume would not come into being.

We thank all the authors for their participation in this project and for their invaluable contribution. A number of colleagues rendered generous help in the preparation of the book; the assistance and dedicated work from a number of anonymous referees are greatly appreciated. Finally, we thank the editor Professor Manfred Thoma and the Springer-Verlag professionals in helping us to finalize the book.

George Yin, Wayne State University
Qing Zhang, University of Georgia

LIST OF AUTHORS

J.R. BIRGE, Industrial and Operations Engineering Department, University of Michigan, 1205 Beal Ave., Ann Arbor, MI 48109,

E.K. BOUKAS, Mechanical Engineering Department, École Polytechnique de Montréal and the GERAD, Montréal, Que., Canada

W.K. CHING, Department of Systems Engineering and Engineering Management, The Chinese University of Hong Kong, Shatin, Hong Kong

J.-Q. HU, Manufacturing Engineering Department, Boston University, 44 Cummington Street, Boston, MA 02215

L. HUANG, Manufacturing Engineering Department, Boston University, 44 Cummington Street, Boston, MA 02215

D. JACOBS, Ford Motor Company, EFHD, McKean and Textile Roads, P.O.Box 922, Ypsilanti, MI 48197

C.-T. KUO, Division of Systems Science, Department of Electrical Engineering and Computer Science, The University of Michigan, Ann Arbor, MI 48109-2122

J.-T. LIM, Division of Systems Science, Department of Electrical Engineering and Computer Science, The University of Michigan, Ann Arbor, MI 48109-2122; on leave from the Department of Electrical Engineering, Korea Advanced Institute of Science and Technology, Taejon, Korea

M.J. MADDOX, Ford Motor Co., Information Systems Planning, Suite 425, 555 Republic Drive, Allen Park, MI 48101

S.M. MEERKOV, Division of Systems Science, Department of Electrical Engineering and Computer Science, The University of Michigan, Ann Arbor, MI 48109-2122

J.R. PERKINS, Manufacturing Engineering Department, Boston University, 44 Cummington Street, Boston, MA 02215

S.P. SETHI, Faculty of Management, University of Toronto, Toronto, Canada M5S 1V4

R. SRIKANT, Coordinated Science Laboratory, University of Illinois, 1308 W. Main St., Urbana, IL 61801

P. VAKILI, Manufacturing Engineering Department, Boston University, 44 Cummington Street, Boston, MA 02215

G. YIN, Department of Mathematics, Wayne State University, Detroit, MI 48202

Q. ZHANG, Department of Mathematics, University of Georgia, Athens, GA 30602

X.Y. ZHOU, Department of Systems Engineering and Engineering Management, The Chinese University of Hong Kong, Shatin, Hong Kong

Q.J. ZHU, Department of Mathematics and Statistics, Western Michigan University, Kalamazoo, MI 49008

Table of Contents

Part I. Optimal Production Planning

1. Matrix Methods in Production Planning of Failure Prone Manufacturing Systems
   W.K. CHING AND X.Y. ZHOU 3

2. Optimal Control of a Failure Prone Manufacturing System with the Possibility of Temporary Increase in Production Capacity
   L. HUANG, J-Q. HU AND P. VAKILI 31

3. Optimal Control of Manufacturing Systems with Buffer Holding Costs
   J.R. PERKINS AND R. SRIKANT 61

Part II. Scheduling and Improvability

4. Using Second Moment Information in Stochastic Scheduling
   M.J. MADDOX AND J.R. BIRGE 99

5. Improvability of Serial Production Lines: Theory and Applications
   D. JACOBS, C.-T. KUO, J.-T. LIM AND M. MEERKOV 121

Part III. Approximate Optimality and Robustness

6. Asymptotic Optimal Feedback Controls in Stochastic Dynamic Two-Machine Flowshops
   S.P. SETHI AND X.Y. ZHOU 147

7. Studying Flexible Manufacturing Systems via Deterministic Control Problems
   Q.J. ZHU 181

8. On Robust Design for a Class of Failure Prone Manufacturing Systems
   E.K. BOUKAS, Q. ZHANG AND G. YIN 197

Part I
Optimal Production Planning

Matrix Methods in Production Planning of Failure Prone Manufacturing Systems[*]

Wai Ki Ching and Xun Yun Zhou

Abstract

This paper studies optimal hedging policy for a failure prone one-machine system. The machine produces one type of product and its demand has batch arrival. The inter-arrival time of the demand, up time of the machine and processing time for one unit of product are exponentially distributed. When the machine is down, it is subject to a sequence of l repairing phases. In each phase, the repair time is exponentially distributed. The machine states and the inventory levels are modeled as Markov processes and an efficient algorithm is presented to solve the steady state probability distribution. The average running cost for the system can be written in terms of the steady state distribution. The optimal hedging point can then be obtained by varying different values of hedging point.

Key Words: Manufacturing System, Hedging Point Policy, Steady State Distribution, Block Gauss-Seidel Method

1 Introduction

In this paper, we study a failure prone manufacturing system. The system consists of one machine producing one type of product. The machine is subject to random breakdowns and repairs. The processing time for one unit of product and the up time of the machine are exponentially distributed. The demand of the product is modeled as batch arrival with the inter-arrival time being exponentially distributed. The batch size distribution is independent of time. If the arrival demand exceeds the maximum allowable backlog, we suggest two possible strategies to handle the issue. We may either accept part

---

[*]This work was partly supported by CUHK Direct Grant No.220500660 and RGC Earmarked Grant CUHK 249/94E.

of the demand (up to the maximum allowable backlog) or reject all the demand. A penalty cost of c_P is assigned to the system for each unit of rejected demand. We name the first strategy as Strategy I and the second strategy as Strategy II.

It is well known that the hedging point policy is optimal for one machine manufacturing systems in some simple situations, see [1, 2, 10, 11, 19]. For two-machine flowshops, hedging policies are no longer optimal but near-optimal and they are studied in [22, 20]. For general flowshops or even jobshops, near-optimal production policies are derived by using the elegant hierarchical control approach [23, 24]. However, it should be noted that in these work, the machine repair time is modeled as exponentially distributed, which is not a reasonable assumption in reality. Very often, instead of one task, a repair requires a sequence of tasks. In this paper, we will consider this general situation for the repair time. When the machine breaks down, it is subjected to a sequence of l repairing phases and in each phase the repair time is exponentially distributed. Moreover, the demand is generalized to be of batch arrival with inter-arrival time being exponentially distributed.

Let us first define the following parameters for the model as follows:

$1/\lambda$:   *the mean up time of the machine,*
$1/d$:   *the mean processing time for 1 unit of product,*
$1/\mu_i$:   *the mean repair time for the machine in phase i, $i = 1, 2, \cdots, l$,*
$1/a$:   *the mean inter $-$ arrival time of demand,*
q_i :   *the probability that the arrival demand requires i units of product, for $i = 1, 2, \cdots, \infty$ and $\sum_{i=1}^{\infty} q_i = 1$,*
m :   *the maximum allowable backlog,*
b :   *the maximum inventory capacity,*
h :   *the hedging point,*
$c_I > 0$:   *the inventory cost per unit of product,*
$c_B > 0$:   *the backlog cost per unit of product,*
$c_P > 0$:   *the penalty cost for rejecting a unit of an arrival demand.*

We focus ourselves in finding the optimal hedging point for the machine-inventory model. Usually, proper positive inventory level is maintained to hedge against the uncertainties in supply, demand or breakdowns of the machine. The hedging point policy is characterized by a number h. The machine keeps on producing products as much as it can if the inventory level is less than h and the machine state is "up", otherwise the machine is shut down or under repair. The machine states and the inventory levels are modeled as Markov processes in some way. It turns out that the combined machine-inventory process is an irreducible continuous time Markov chain. We give the generator for the process, and then use the Block Gauss-Seidel (BGS) method to compute the steady state probability distribution. The average running cost for the system can then be written in terms of the steady state probability distribution, and therefore the optimal hedging point can be obtained by varying different values of h.

The remainder of this paper is organized as follows. In §2, we formulate the machine-inventory model, and write down the generator of the corresponding Markov chain. We then write the average running cost functions for both Strategy I and Strategy II. Our objective is to find the optimal hedging point h which minimizes the average running cost functions. In §3, we present an algorithm to solve the steady state probability distribution by using the BGS method. In addition, we prove the convergence of the BGS method. In §4, we discuss a fast algorithm for solving the steady state probability distribution when the batch size is restricted to 1. In §5, numerical examples are given to demonstrate the fast convergent rate of our methods. Concluding remarks are given in §6 to address some possible generalizations of our model.

2 The Machine-inventory Model

Let $\alpha(t)$ be the state of the machine at time t given by

$$\alpha(t) = \begin{cases} 0, & \text{if the machine is up,} \\ i, & \text{if the machine is under repairing in phase } i, \\ & i = 1, \cdots, l, \end{cases}$$

and $x(t)$ be the inventory level at time t which takes integer values in $[-m, h]$. We assume that the initial inventory level is zero without loss of generality. Then the machine-inventory process $\{(\alpha(t), x(t)), t \geq 0\}$ is a continuous time

Markov chain taking values in the state space

$$S = \{(\alpha, x) : \alpha = 0, \cdots, l, \ x = -m, \cdots, h.\}.$$

Each time when visiting a state, the process stays there for a random period of time that has an exponential distribution and is independent of the past behavior of the process. Let us derive the generator for the machine-inventory process. The machine process follows an irreducible Markov chain. If we order the machine state space in ascending order of phases of repairing with the last state being the up state, then we can obtain the following $(l+1) \times (l+1)$ generator M for the machine process.

$$M = \begin{bmatrix} \mu_1 & & & & -\lambda \\ -\mu_1 & \mu_2 & & & \\ & -\mu_2 & \ddots & & \\ & & \ddots & \mu_l & \\ & & & -\mu_l & \lambda \end{bmatrix}. \quad (1)$$

The generator M has positive diagonal entries, non-positive off-diagonal entries and each column sum of M is zero. It should be noted that the generator could have been defined in the usual way as

$$\tilde{M} = \begin{bmatrix} -\mu_1 & \mu_1 & & & \\ & -\mu_2 & \mu_2 & & \\ & & \ddots & \ddots & \\ & & & -\mu_l & \mu_l \\ \lambda & & & & -\lambda \end{bmatrix}. \quad (2)$$

Here we redefined our generator to be $M = -\tilde{M}^t$ just for the convenience in discussion. The inventory level follows a continuous time Markov chain. If we order the inventory state space in descending order, the generators for the inventory levels under Strategy I and Strategy II when the machine is in the "up" state are given by the following $(m+h+1) \times (m+h+1)$ matrices T_{11}

and T_{21}, respectively, where

$$T_{11} = \begin{bmatrix} a & -d & & & & & \\ -a_1 & a+d & -d & & & & \\ -a_2 & -a_1 & \ddots & & -d & & \\ \vdots & \ddots & \ddots & \ddots & \ddots & & \\ -a_{m+h-1} & & \vdots & -a_2 & -a_1 & a+d & -d \\ -\gamma_{m+h} & & \cdots & \cdots & -\gamma_2 & -\gamma_1 & d \end{bmatrix}, \qquad (3)$$

and

$$T_{21} = \begin{bmatrix} \beta_{m+h} & & -d & & & & \\ -a_1 & \beta_{m+h-1}+d & -d & & & & \\ -a_2 & & -a_1 & \ddots & \ddots & & \\ \vdots & & -a_2 & \ddots & \ddots & \ddots & \\ \vdots & & & \vdots & \ddots & \ddots & \beta_1+d & -d \\ -a_{m+h} & & \cdots & \cdots & -a_2 & -a_1 & d \end{bmatrix}, \qquad (4)$$

where
$$\begin{cases} a_i = aq_i, \\ \gamma_i = \sum_{k=i}^{\infty} a_k, \\ \beta_i = \sum_{k=1}^{i} a_k. \end{cases}$$

This kind of generators occurs in many applications, such as queueing systems, communication systems and loading dock models, see [5, 21, 18] for instance. If the machine remains in the repairing states, then the generators for the inventory levels under Strategy I and Strategy II are given by the following $(m+h+1) \times (m+h+1)$ lower triangular matrices T_{10} and T_{20}, respectively, where

$$T_{10} = \begin{bmatrix} a & & & & & & \\ -a_1 & a & & & & & \\ -a_2 & -a_1 & \ddots & & & & \\ \vdots & -a_2 & \ddots & \ddots & & & \\ -a_{m+h-1} & & \vdots & \ddots & -a_1 & a & \\ -\gamma_{m+h} & & \cdots & \cdots & -\gamma_2 & -\gamma_1 & 0 \end{bmatrix}, \qquad (5)$$

and

$$T_{20} = \begin{bmatrix} \beta_{m+h} & & & & & \\ -a_1 & \beta_{m+h-1} & & & & \\ -a_2 & -a_1 & \ddots & & & \\ \vdots & -a_2 & \ddots & \ddots & & \\ \vdots & \vdots & \ddots & \ddots & \beta_1 & \\ -a_{m+h} & \cdots & \cdots & -a_2 & -a_1 & 0 \end{bmatrix}. \quad (6)$$

Therefore the generators for the combined machine-inventory process under Strategy I and Strategy II are given by the following $[(l+1)(m+h+1)] \times [(l+1)(m+h+1)]$ matrices $A_{i,m,h}, i = 1, 2$, respectively, where

$$A_{i,m,h} = \begin{bmatrix} T_{i0} + \mu_1 I & & & & -\lambda I \\ -\mu_1 I & T_{i0} + \mu_2 I & & & \\ & -\mu_2 I & \ddots & & \\ & & \ddots & T_{i0} + \mu_l I & \\ & & & -\mu_l I & T_{i1} + \lambda I \end{bmatrix}. \quad (7)$$

We are interested in the steady state of the system. Let

$$p(\alpha, x) = \lim_{t \to \infty} \text{Prob}(\alpha(t) = \alpha, x(t) = x)$$

be the steady state probability distribution. Let

$$p(i) = \sum_{k=0}^{l} p(k, i), \quad i = -m, -(m-1), \cdots, 0, \cdots, h$$

be the steady state probability of the inventory level of the system. The average running cost is of the following form :

Running cost = Inventory Cost + Backlog Cost + Penalty Cost

In Strategy I, the system accepts arrival demand only up to the maximum allowable backlog. The average running cost for the machine-inventory system under Strategy I is given as follow:

$$C_1(h) = c_I \sum_{i=1}^{h} i p(i) - c_B \sum_{i=-m}^{-1} i p(i) + c_P \sum_{i=-m}^{h} [p(i) \sum_{k=m+i+1}^{\infty} (k-m-i)q_k].$$

In Strategy II, the system rejects the whole batch of demand if accepting the demand would lead to a backlog more than m. Therefore the average cost for the machine-inventory system under Strategy II is given as follow:

$$C_2(h) = c_I \sum_{i=1}^{h} ip(i) - c_B \sum_{i=-m}^{-1} ip(i) + c_P \sum_{i=-m}^{h} [p(i) \sum_{k=m+i+1}^{\infty} kq_k].$$

Our main objective is to determine h which minimizes the average running cost functions $C_1(h)$ under Strategy I and $C_2(h)$ under Strategy II. In the next section, we will discuss iterative methods for computing the steady state probability distribution $p(\alpha, x)$. For ease of presentation, we will consider Strategy I only in the following discussion and therefore denote $A_{m,h} = A_{1,m,h}$.

3 Iterative Methods for Solving the Steady State Distribution

In this section, we discuss numerical methods for solving the steady state distribution $p(\alpha, x)$, which is the right positive null vector of the matrix $A_{m,h}$. We shall make use of the block structure of the generator $A_{m,h}$ to save computational cost and memory. To this end, we employ the Block Gauss-Seidel (BGS) method (see Varga [27]), because it is easy to implement and has stable convergence rate. We prove that BGS converges and construct an initial guess which is close to the solution $p(\alpha, x)$. The convergence rate of BGS can be improved by adjusting this initial guess properly. The main computational cost for BGS comes from solving the linear diagonal block systems of $A_{m,h}$. We develop a fast algorithm to solve this issue in $O(l(m+h)\log(m+h))$ operations. The computational cost and memory requirement are also discussed.

Now let us first show the existence of the steady state probability distribution. Note that the generator $A_{m,h}$ is irreducible, has zero column sum, positive diagonal entries and non-positive off-diagonal entries, so $(I - A_{m,h} \cdot (\text{diag}(A_{m,h}))^{-1})$ is a stochastic matrix. Here $\text{diag}(A)$ is the diagonal matrix containing the diagonal entries of the matrix A. From the Perron and Frobenius theory, we know that $A_{m,h}$ has a one-dimensional null-space with a right positive null vector, see Varga [27, p. 30]. The steady state probability distribution is then equal to the normalized form of the positive null vector. Since

$A_{m,h}$ is singular, we consider an equivalent linear system

$$G_{m,h}\mathbf{x} \equiv (A_{m,h} + \mathbf{f_n f_n}^t)\mathbf{x} = \mathbf{f_n}, \qquad (8)$$

where $\mathbf{f_n} = (0, \cdots, 0, \delta)^t$ is an $(l+1)(m+h+1)$ column vector with some $\delta > 0$. The following lemma shows that the linear system (8) is non-singular and hence the steady state probability distribution can be obtained by normalizing the solution of equation (8), see [14].

Lemma 1 *The matrix $G_{m,h}$ is non-singular.*

Proof Since the matrix $G_{m,h}$ is an irreducible matrix, by applying the Gerschgorin circle theorem (see Horn [9, p.346]) to the matrix $G_{m,h}$, we see that all the eigenvalues has non-negative real part. Moreover, by applying Theorem 1.7 in Varga [27, p.20] to the matrix $G_{m,h}$, we know that zero cannot be an eigenvalue of $G_{m,h}$. Thus the matrix $G_{m,h}$ is non-singular. □

3.1 Convergence of the Block Gauss-Seidel Method

In this section, we prove that the classical BGS method converges when applied to solving the linear system (8). For the algorithm of BGS, see Appendix and [8].

We begin with the following lemma.

Lemma 2 *In solving the linear system (8), the Block Jacobi method converges if and only if the Block Gauss-Seidel method converges.*

Proof We note that $G_{m,h}$ is a consistently ordered $(l+1)$-cyclic matrix, our result follows from [27, p. 107]. □

It then suffices to prove the convergence of the Block Jacobi method.

Theorem 1 *The Block Jacobi method converges.*

Proof We note that both $G_{m,h}$ and $(T_{11} + \lambda I + \delta \mathbf{e_n e_n}^t)$ are real, non-singular and irreducible diagonally dominant matrices, where $\mathbf{e_n} = (0, \cdots, 0, 1)^t$ is an

$(m+h+1)$ column vector. They have positive diagonal entries and non-positive off-diagonal entries. Therefore $G_{m,h}^{-1}$ and $(T_{11} + \lambda I + \delta e_n e_n^t)^{-1}$ are non-negative matrices, see Varga [27, p. 85]. Moreover, one can show that the matrices $(T_{10} + \mu_i I)^{-1}, (i = 1, 2, \cdots, l)$ are also non-negative matrices by using the fact that $(T_{10} + \mu_i I)$ is non-singular M-matrices. Let $D_{m,h}$ be the matrix containing the diagonal blocks of the matrix $G_{m,h}$. Then

$$D_{m,h} - (D_{m,h} - G_{m,h})$$

is a regular spitting of the matrix $G_{m,h}$. It follows that the spectral radius

$$\rho(D_{m,h}^{-1}(D_{m,h} - G_{m,h})) < 1$$

by Theorem 3.13 in Varga [27, p. 89]. Hence the Block Jacobi method converges. □

3.2 Fast Solution for the Diagonal Blocks Systems

The main computational cost for the BGS method comes from the computation of the inverses of the dense diagonal blocks in order to solve the following linear systems

$$(T_{10} + \mu_i I)\mathbf{y} = \mathbf{b}, \; i = 1, \cdots, l, \quad \text{and} \quad (T_{11} + \lambda I + \mathbf{e_n e_n}^t)\mathbf{y} = \mathbf{b} \quad (9)$$

for any given right hand side vector **b**. We remark that the linear systems (9) can be solved in $O((m+h)^2)$ operations by Gaussian elimination with $O((m+h)^2)$ memory, see Chan and Ching [5]. In the following, we present an iterative method for solving each of the linear systems of (9) in $O((m+h)\log(m+h))$ operations with $O(m+h)$ memory. The algorithm is based on the Conjugate Gradient Squared (CGS) method (see Sonneveld [25]). The convergence rate of the Conjugate Gradient (CG) method when applied to solve a linear system $Q\mathbf{y} = \mathbf{b}$ depends on the singular values of Q. The more clustered the singular values of Q are around one, the faster the convergence rate will be. This method usually uses a preconditioner to accelerate the convergence rate, so it is also called the Preconditioned Conjugate Gradient Squared (PCGS) method, see Golub [8]. A good preconditioner P should satisfy the following :

(i) the preconditioned system $P^{-1}Q$ has clustered singular values around one;

(ii) the inverse P^{-1} is easy to compute; and

(iii) the preconditioner P^{-1} is easy to construct.

In the PCGS method, we solve the linear system $P^{-1}Q\mathbf{y} = P^{-1}\mathbf{b}$ instead of $Q\mathbf{y} = \mathbf{b}$. We construct preconditioners by using the near-Toeplitz like structure of T_{10} and T_{11}, and prove that the preconditioned systems have singular values clustered around 1. Let us first give the definition of Toeplitz matrices and some of its properties.

Definition 1 *An $n \times n$ matrix $S_n = [s_{i,j}]$ is said to be Toeplitz if $s_{i,j} = s_{i-j}$, i.e., if S_n is constant along all its diagonals. It is said to be circulant if its diagonals s_k further satisfies $s_{n-k} = s_{-k}$ for $0 < k \leq n - 1$.*

Definition 2 *Let f be a 2π-periodic continuous complex-valued function defined on the interval $[-\pi, \pi]$, and*

$$s_k = \frac{1}{2\pi} \int_{-\pi}^{\pi} f(\theta) e^{-ik\theta} d\theta, \quad k = 0, \pm 1, \pm 2, \pm 3, \cdots.$$

The function f is called the generating function of the sequence of $n \times n$ Toeplitz matrices $S_n[f]$, if $S_n[f]_{j,k} = s_{j-k}$, for $0 \leq j, k < n$.

We note that any $n \times n$ circulant matrix C_n is characterized by its first column and can be diagonalized by the discrete Fourier matrix F_n, i.e., $C_n = F_n^* \Lambda_n F_n$, where the entries of F_n are given by

$$[F_n]_{j,k} = \frac{1}{n} e^{\frac{2\pi ijk}{n}},$$

F_n^* is the conjugate transpose of F_n and Λ_n is a diagonal matrix containing the eigenvalues of C_n, see Davis [7, p. 74]. The matrix-vector multiplication of the forms $F_n\mathbf{y}$ and $F_n^*\mathbf{y}$ can be obtained in $O(n\log n)$ operations by the Fast Fourier Transform (FFT). Moreover, by using the following equations

$$C_n^1 = C_n\mathbf{e_1} = F_n^*\Lambda_n F_n\mathbf{e_1} = F_n^*\Lambda_n\mathbf{1} = F_n^*\Lambda_n^1,$$

the eigenvalues of C_n can be obtained in $O(n\log n)$ operations. Here $\mathbf{e_1} = (1,0,\cdots,0)^t$, C_n^1 is the first column of C_n, $\mathbf{1} = (1,\cdots,1)^t$ and Λ_n^1 is the column vector containing all the eigenvalues of C_n. Hence solving a circulant linear system $C_n\mathbf{y} = \mathbf{b}$ requires only $O(n\log n)$ operations. We further remark that, for a Toeplitz matrix S_n, the matrix-vector multiplication of the form $S_n\mathbf{y}$ can be done in $O(n\log n)$ operations by embedding S_n into a $2n \times 2n$ circulant matrix, see Strang [26] for details.

Note that the idea of using circulant matrices as preconditioners for solving Toeplitz systems has been studied extensively in recent years, see for instance [26, 3, 16]. For a given Toeplitz matrix S_n with diagonals $\{s_j\}_{j=-(n-1)}^{n-1}$, a circulant preconditioner to S_n is defined to be the circulant matrix C_n which minimizes the Frobenius norm $||S_n - C_n||_F$ over the set of all circulant matrices. The (i,j)th entry of the matrix C_n is given by c_{i-j} where

$$c_k = \begin{cases} \dfrac{(n-k)s_k + ks_{k-n}}{n}, & 0 \leq k < n, \\ c_{n+k}, & 0 < -k < n, \end{cases} \qquad (10)$$

see Chan [3]. For a sequence of Toeplitz matrices $T_n[f]$ with generating function f, we denote Chan's circulant matrices as $C_n[f]$.

In the following, we construct circulant preconditioners for the matrices $(T_{10} + \mu_i I_n)$ (for $i = 1,\cdots,l$) and $(T_{11} + \lambda I_n + \mathbf{e_n}\mathbf{e_n}^t)$. Here I_n is the $n \times n$ identity matrix. Let us first define the following Toeplitz matrices,

$$T^0 = \begin{bmatrix} \psi & -d & & & & \\ -a_1 & \ddots & \ddots & & & \\ -a_2 & \ddots & \ddots & \ddots & & \\ \vdots & -a_2 & -a_1 & \psi & -d \\ \cdots & & -a_2 & -a_1 & \psi \end{bmatrix}, \qquad (11)$$

and

$$T^i = \begin{bmatrix} a+\mu_i & & & & \\ -a_1 & \ddots & & & \\ -a_2 & \ddots & \ddots & & \\ \vdots & -a_2 & -a_1 & a+\mu_i & \\ \cdots & & -a_2 & -a_1 & a+\mu_i \end{bmatrix}, \qquad (12)$$

where $\psi = a + d + \lambda$ and $i = 1, 2, \cdots, l$. By direct verification we have the following lemma concerning the generating functions for T^i and T^0.

Lemma 3 *The generating functions for the Toeplitz matrices T^0 and T^i, $i = 1, \cdots, l$, are given by*

$$g_0(\theta) = (a+d+\lambda) - de^{i\theta} - \sum_{k=1}^{\infty} a_k e^{-ik\theta}$$

and

$$g_i(\theta) = (a+\mu_i) - \sum_{k=1}^{\infty} a_k e^{-ik\theta}$$

respectively. Furthermore,

$$\text{rank}(T_{11} + \lambda I + \delta \mathbf{e_n}\mathbf{e_n}^t - T^0) = 2 \text{ and } \text{rank}(T_{10} + \mu_i I - T^i) = 1.$$

We define the circulant preconditioners for the matrices $(T_{11}+\lambda I_n+\delta \mathbf{e_n}\mathbf{e_n}^t)$ and $(T_{10} + \mu_i I_n)$ to be $C_n[g_0]$ and $C_n[g_i]$, respectively. The following theorem shows that the preconditioned systems have singular values clustered around 1, hence the PCGS method applied to solve the preconditioned systems will converge very fast.

Theorem 2 *The preconditioned systems*

$$C_n[g_0]^{-1}(T_{11} + \lambda I + \delta \mathbf{e_n}\mathbf{e_n}^t) \qquad (13)$$

and

$$C_n[g_i]^{-1}(T_{10} + \mu_i I_n), \quad i = 1, 2, \cdots, l, \qquad (14)$$

have their singular values clustered around one.

Proof Here we prove the theorem for system (13) only. The proof for system (14) is similar and therefore omitted. Note that

$$C_n[g_0]^{-1}(T_{11} + \lambda I + \delta \mathbf{e_n}\mathbf{e_n}^t) = C_n[g_0]^{-1}T^0 + L,$$

where $L = C_n[g_0]^{-1}(T_{11} + \lambda I + \delta \mathbf{e_n}\mathbf{e_n}^t - T^0)$, and $\text{rank}(L) = 2$ by Lemma 3. By considering the real part of $g_0(\theta)$, the generating function g_0 does not vanish on the unit circle of the complex plane. Moreover g_0 is a 2π-periodic continuous function. By Chan and Yeung [6, Corollary 1], we have

$$C_n[g_0]^{-1}T^0 = I + L_n + U_n,$$

where L_n is a low rank matrix and U_n is a small l_2-norm matrix. It follows that

$$C_n[g_0]^{-1}(T_{11} + \lambda I + \delta \mathbf{e_n}\mathbf{e_n}^t) = I + (L_n + L) + U_n. \tag{15}$$

By applying the Cauchy interlace theorem (Wilkinson [28, p. 103]) to the matrix system (15), we have the singular values of $C_n[g_0]^{-1}(T_{11} + \lambda I + \mathbf{e_n}\mathbf{e_n}^t)$ clustered around one. □

By Theorem 2, the PCGS method will converge superlinearly when applied to solve the preconditioned linear systems. For the numerical examples and convergence behavior, see [6] and [5] for more details. We remark that when applying the same method to solve the linear systems

$$(T_{20} + \mu_i I)\mathbf{y} = \mathbf{b}, \; i = 1, \cdots, l \quad \text{and} \quad (T_{21} + \lambda I + \delta \mathbf{e_n}\mathbf{e_n}^t)\mathbf{y} = \mathbf{b} \tag{16}$$

for any right handside vector **b**, we also get a very fast convergence rate, see [5] for example.

3.3 Computational Cost and Memory Requirement

In applying the PCGS method to solve the preconditioned system $P^{-1}Q\mathbf{y} = \mathbf{b}$, the main computational cost is due to the vector-matrix multiplication of $Q\mathbf{y}$ and solving the preconditioner system $P\mathbf{y} = \mathbf{w}$. Our preconditioner P is of circulant type, therefore can be inverted in $O((m+h)\log(m+h))$ operations.

By exploiting the near-Toeplitz structure of the matrices T_{10}, T_{11}, T_{20} and T_{21}, one can show the following lemma regarding matrix-vector multiplication of $T_{i0}\mathbf{y}$ and $T_{i1}\mathbf{y}$, $i = 1, 2$.

Lemma 4 *The vector-matrix multiplication of $T_{i0}\mathbf{y}$ and $T_{i1}\mathbf{y}$, $i = 1, 2$, can be done in $O((m+h)\log(m+h))$ operations.*

Proof We note that T_{11} and T_{10} differ from a Toeplitz matrix by their first and last row. We can make use of this property to compute $T_{11}\mathbf{y}$ and $T_{10}\mathbf{y}$ in $O((m+h)\log(m+h))$ operations. The matrices T_{21} and T_{20} differ from a Toeplitz matrix by their main diagonals only. Therefore the matrix-vector multiplication $T_{21}\mathbf{y}$ and $T_{20}\mathbf{y}$ can also be done in $O((m+h)\log(m+h))$ operations. □

Hence in each iteration of the PCGS method, the operation cost is $O((m+h)\log(m+h))$. We have proved in Theorem 2 that the preconditioned system has singular values clustered around one, so the PCGS method will converge very fast independent of the matrix size $(m+h)$. Thus the cost for solving $T_{i0}\mathbf{y} = \mathbf{b}$ and $T_{i1}\mathbf{y} = \mathbf{b}$ is of the order $O((m+h)\log(m+h))$. We have the following lemma concerning the cost per iteration of the BGS.

Lemma 5 *The operation cost per iteration for the BGS method is $O(l(m+h)\log(m+h))$.*

For the memory requirement of our method, we need to store the approximated solution obtained in each iteration step which will require $O(l(m+h))$ memory. We do not need to store all the diagonal blocks matrices T_{i0} and T_{i1}. By using the near-Toeplitz structure of the matrices, and the fact that any Toeplitz matrix is identified by its first row and column, we only need $O(l(m+h))$ memory, see [5]. Moreover, we need another $O(l(m+h))$ memory to store the eigenvalues of the circulant preconditioners. Hence the total memory cost for our method is $O(l(m+h))$.

4 A Special Case

In this section, we discuss a fast iterative algorithm for solving the steady state probability distribution when the arrival batch size is one. Once again, we solve the non-singular equation (8) by the PCGS method.

4.1 Construction of Preconditioners

In this section, we construct preconditioners by taking what we call "circulant approximation" of the generator $A_{m,h}$. First note that when the batch size is one, the generator becomes

$$A_{m,h} = \begin{bmatrix} D_1 & & & & -\lambda I \\ -\mu_1 I & D_2 & & & \\ & -\mu_2 I & \ddots & & \\ & & \ddots & D_l & \\ & & & -\mu_l I & D_{l+1} \end{bmatrix}, \qquad (17)$$

where

$$D_k = \begin{bmatrix} a+\mu_k & & & & \\ -a & a+\mu_k & & & \\ & -a & \ddots & & \\ & & \ddots & a+\mu_k & \\ & & & -a & \mu_k \end{bmatrix}, \quad k=1,\cdots,l, \qquad (18)$$

and

$$D_{l+1} = \begin{bmatrix} a+\lambda & -d & & & \\ -a & a+d+\lambda & \ddots & & \\ & -a & \ddots & & -d \\ & & \ddots & a+d+\lambda & -d \\ & & & -a & d+\lambda \end{bmatrix}. \qquad (19)$$

We define the circulant approximation $c(D_k)$ for D_k as follow:

$$c(D_k) = \begin{bmatrix} a+\mu_k & & & & -a \\ -a & a+\mu_k & & & \\ & -a & \ddots & & \\ & & \ddots & a+\mu_k & \\ & & & -a & a+\mu_k \end{bmatrix}, \quad k=1,\cdots,l, \qquad (20)$$

and

$$c(D_{l+1}) = \begin{bmatrix} \psi & -d & & & & -a \\ -a & \psi & \ddots & & & \\ & -a & \ddots & -d & & \\ & & \ddots & & \psi & -d \\ -d & & & & -a & \psi \end{bmatrix}, \quad (21)$$

where $\psi = a + d + \lambda$. By direct verification, we have the following lemma 6, which indicates that the circulant approximation of D_i is close to itself by at most a matrix of rank two.

Lemma 6 *We have* rank $(D_k - c(D_k)) = 1$, *for* $k = 1, \cdots, l$, *and* rank $(D_{l+1} - c(D_{l+1})) = 2$.

The following matrix A_0 is obtained by taking the circulant approximation of the diagonal blocks of $A_{m,h}$:

$$A_0 = \begin{bmatrix} c(D_1) & & & & -\lambda I \\ -\mu_1 I & c(D_2) & & & \\ & -\mu_2 I & \ddots & & \\ & & \ddots & c(D_l) & \\ & & & -\mu_l I & c(D_{l+1}) \end{bmatrix}. \quad (22)$$

We note that A_0 has zero-sum columns, positive diagonal entries and non-positive off-diagonal entries. Therefore A_0 is singular and has one dimensional null space. Moreover, each circulant diagonal block $c(D_i)$ of A_0 can be diagonalized by the discrete Fourier transform F_N, where $N = m + h$. Namely

$$F_N c(D_k) F_N^* = \begin{bmatrix} \theta_{k,1} & & & \\ & \theta_{k,2} & & \\ & & \ddots & \\ & & & \theta_{k,N} \end{bmatrix}, \quad k = 1, \cdots, l, \quad (23)$$

and

$$F_N c(D_{l+1}) F_N^* = \begin{bmatrix} \theta_{l+1,1} & & & \\ & \theta_{l+1,2} & & \\ & & \ddots & \\ & & & \theta_{l+1,N} \end{bmatrix}. \quad (24)$$

The eigenvalues are given by

$$\theta_{k,j} = \mu_k + a(1 - e^{\frac{-2\pi i(j-1)}{N}}), \quad k = 1, \cdots, l, \text{ and } j = 1, \cdots, N,$$

and

$$\theta_{l+1,j} = \lambda + a(1 - e^{\frac{-2\pi i(j-1)}{N}}) + d(1 - e^{\frac{-2\pi i(N-1)(j-1)}{N}}), \quad j = 1, \cdots, N,$$

see Davis [7, p. 14]. Therefore we have the matrix

$$(I_N \otimes F_N) A_0 (I_N \otimes F_N^*)$$
$$= \begin{bmatrix} F_N c(D_1) F_N^* & & & & -\lambda I_N \\ -\mu_1 I_N & F_N c(D_2) F_N^* & & & \\ & -\mu_2 I_N & \ddots & & \\ & & \ddots & F_N c(D_l) F_N^* & \\ & & & -\mu_l I_N & F_N c(D_{l+1}) F_N^* \end{bmatrix}, \quad (25)$$

where I_N is the $N \times N$ identity matrix, and $\otimes$ is the Kronecker tensor product. There exists a permutation matrix P such that

$$P(I_N \otimes F_N) A_0 (I_N \otimes F_N^*) P^t = \begin{bmatrix} E_1 & & & & \\ & E_2 & & & \\ & & \ddots & & \\ & & & E_{N-1} & \\ & & & & E_N \end{bmatrix}, \quad (26)$$

where

$$E_k = \begin{bmatrix} \omega_{k,1} & & & & -\lambda \\ -\mu_1 & \omega_{k,2} & & & \\ & -\mu_2 & \ddots & & \\ & & \ddots & \omega_{k,l} & \\ & & & -\mu_l & \omega_{k,l+1} \end{bmatrix}, \quad k = 1, \cdots, l+1, \quad (27)$$

and $\omega_{k,j}$ are given by

$$\omega_{k,j} = \mu_j + a(1 - e^{\frac{-2\pi i(k-1)}{N}}), \quad \text{for } k = 1, \cdots, l, \text{ and } j = 1, \cdots, N,$$

and

$$\omega_{l+1,j} = \lambda + a(1 - e^{\frac{-2\pi i(k-1)}{N}}) + d(1 - e^{\frac{-2\pi i(N-1)(j-1)}{N}}), \quad \text{for } j = 1, \cdots, N.$$

We note that $E_1 \equiv M$, the generator for the machine process. Therefore E_1 is the only singular block, because M is singular and A_0 has one dimension null space. We define $\tilde{E}_1 = (E_1 + \mathbf{e}_{l+1}\mathbf{e}_{l+1}{}^t)$, where $\mathbf{e}_{l+1} = (0, \cdots, 0, 1)^t$ is an $(l+1)$ unit column vector. By similar argument to that in the proof of Lemma 1, one can show that the matrix $\tilde{E}_1$ is non-singular. Define

$$\tilde{E} = \begin{bmatrix} \tilde{E}_1 & & & & \\ & E_2 & & & \\ & & \ddots & & \\ & & & E_{N-1} & \\ & & & & E_N \end{bmatrix}. \qquad (28)$$

Our preconditioner is defined as

$$\tilde{A}_0 = (I_N \otimes F_N^*) P^t \tilde{E} P (I_N \otimes F_N),$$

which is clearly non-singular.

4.2 Convergence Analysis

In this section, we prove that the preconditioned system $\tilde{A}_0^{-1} G_{m,h}$ has clustered singular values around one.

Theorem 3 *The preconditioned system $\tilde{A}_0^{-1} G_{m,h}$ has singular values clustered around one.*

Proof We first note that from the definitions of A_0 and $\tilde{A}_0$, it follows that

$$\mathrm{rank}(G_{m,h} - A_0) = l + 1 \quad \text{and} \quad \mathrm{rank}(A_0 - \tilde{A}_0) = 1.$$

Therefore we have

$$\mathrm{rank}(G_{m,h} - \tilde{A}_0) \leq \mathrm{rank}(G_{m,h} - A_0) + \mathrm{rank}(A_0 - \tilde{A}_0) = l + 2.$$

Hence

$$\tilde{A}_0^{-1} G_{m,h} = I + \tilde{A}_0^{-1}(G_{m,h} - \tilde{A}_0) \equiv I + L, \qquad (29)$$

where $\mathrm{rank}(L) = l+2$. By applying Cauchy interlace theorem to (29) (Wilkinson [28, p.103]), we conclude that the singular values of $\tilde{A}_0^{-1} G_{m,h}$ are clustered around one. □

4.3 Computational Cost and Memory Requirement

The computational cost of the PCGS method comes from the matrix-vector multiplication $G_{m,h}\mathbf{x}$ as well as solving the preconditioner system. For the first issue, we need $O(l(m+h))$ operations. Let us now discuss the cost for solving the preconditioner system

$$\tilde{A}_0 \mathbf{x} = \mathbf{b}. \tag{30}$$

Solving equation (30) is equivalent to solving sequentially the following equations

$$\tilde{E}\mathbf{y} = P(I \otimes F_N)\mathbf{b} \tag{31}$$

and

$$P(I_N \otimes F_N)\mathbf{x} = \mathbf{y}. \tag{32}$$

The multiplication of the form $(I \otimes F_N)\mathbf{x}$ and $(I \otimes F_N^*)\mathbf{x}$ can be done in $O(l(m+h)\log(m+h))$ operations. The multiplication of the form $P\mathbf{x}$ and $P^t\mathbf{x}$ can be done in $O(l(m+h))$ operations. Therefore solving equation (32) requires $O(l(m+h)\log(m+h))$ operations. Let us consider the cost of solving the diagonal block systems (31). We first prove the following lemma.

Lemma 7 *The linear system* $H\mathbf{x} = \mathbf{b}$ *can be solved in* $O(l)$ *operations if* $\det(H) \neq 0$,

$$H = \begin{bmatrix} a_1 & & & & -c \\ -b_1 & a_2 & & & \\ & -b_2 & \ddots & & \\ & & \ddots & a_l & \\ & & & -b_l & a_{l+1} \end{bmatrix}. \tag{33}$$

Proof Let L be the lower triangular part of the matrix H. We have

$$H = L + (1, 0, \cdots, 0)^t (0, \cdots, 0, -c) \equiv L + \mathbf{u}\mathbf{v}^t.$$

By using the Sherman-Morrison-Woodbury's formula (see Golub [8, p. 51]), the inverse of H can be obtained as

$$H^{-1} = L^{-1}(I - \mathbf{u}(1 + \mathbf{v}^t L^{-1}\mathbf{u})^{-1}\mathbf{v}^t L^{-1}), \tag{34}$$

provided that
$$\Delta = 1 + \mathbf{v}^t H^{-1}\mathbf{u} \neq 0.$$
By direct verification,
$$\Delta = 1 - \frac{b_1 b_2 \cdots b_l c}{a_1 a_2 \cdots a_{l+1}},$$
and
$$\det(H) = a_1 a_2 \cdots a_{l+1} - b_1 b_2 \cdots b_l c.$$

Therefore $\Delta \neq 0$ if and only if $\det(H) \neq 0$. Hence we may solve the linear system $H\mathbf{x} = \mathbf{b}$ by using the formula (34). Clearly, we need only $O(l)$ operations to obtain the solution. □

By lemma 7, we see that solving the diagonal system (31) requires only $O(l(m+h))$ operations. Hence we need $O(l(m+h)\log(m+h))$ computational cost for PCGS method in each iteration. Since we have proved that the PCGS method converges very fast independent of the parameters $N = m + h$, we need approximately $O(l(m+h)\log(m+h))$ operations to obtain the steady state probability distribution. Furthermore, we need $O(l(m+h))$ memory for our method.

Remark 1 Due to the special structure of our preconditioner, we can save more computational cost by using a parallel computer. Usually the parameter l is much less than $N = m + h$. If we have a parallel computer of $l + 1$ processors, then the operation cost for our method can be further reduced to $O((m+h)\log(m+h))$, as the multiplication of $(I \otimes F_N)\mathbf{x}$ and $(I \otimes F_N^*)\mathbf{x}$ can be done by the $l+1$ processors separately and simultaneously. Moreover, the diagonal block system $\tilde{E}\mathbf{x} = \mathbf{b}$ can be solved parallelly to reduce computational cost.

Remark 2 The concept of "circulant approximation" in the construction of preconditioners can be extended to more general cases of the generators. For example we may first approximate the Toeplitz-like diagonal blocks of $A_{m,h}$ by the Toeplitz matrices T^0 and $T^i (i = 1, \cdots, l)$ in (11) and (12), and then approximate the Toeplitz matrices by Chan's circulant matrix to obtain a block circulant preconditioner. One can show that the preconditioner constructed in this way is non-singular in our case. We expect that the PCGS method converges very fast when our preconditioner is used.

5 Numerical Examples

In this section, we illustrate the fast convergence rate of the Block Gauss-Seidel method by some numerical examples. Usually, the convergence rate of BGS depend on both the value of δ in equation (8) and the initial guess, see [14, 4, 15]. The smaller the value of δ the faster the convergence rate of BGS will be. In all the numerical examples below, we set $\delta = 10^{-12}$ and the stopping criteria to be
$$\frac{||A_{m,k}\mathbf{p}^n||_2}{||A_{m,k}\mathbf{p}^0||_2} < 10^{-10},$$
where $\mathbf{p}^n$ is the approximated solution in the n-th iteration and $||\mathbf{y}||_2$ is the 2-norm of a vector $\mathbf{y}$. Here we construct an initial guess by using the steady state probability distributions of the machine states and the inventory states. Let $\mathbf{p_m}, \mathbf{p_d}$ and $\mathbf{p_u}$ be the steady state distribution of the machine states, inventory levels when the machine is down, and inventory levels when the machine is up, respectively. Then by direct verification, we have
$$\mathbf{p_m} = \frac{1}{(\frac{1}{\lambda} + \sum_{i=1}^{l} \frac{1}{\mu_i})}(\frac{1}{\mu_1}, \cdots, \frac{1}{\mu_l}, \frac{1}{\lambda})^t$$
and $\mathbf{p_d} = (0, \cdots, 0, 1)^t$. The probability vector $\mathbf{p_u}$ can be obtained in $O((m+h)\log(m+h))$ operations, see Chan and Ching [5]. Then our initial guess is chosen to be
$$\mathbf{p}^0 = \mathbf{p_m} \otimes (\mathbf{p_d}, \cdots, \mathbf{p_d}, \mathbf{p_u})^t,$$
where $\otimes$ is the Kronecker tensor product. We need $O((m+h)\log(m+h))$ for the construction of $\mathbf{p}^0$.

In our numerical example, we assume that the system runs for 52 weeks (1 year). The mean up time of the machine is 50 weeks, i.e. $\lambda = 1/50$. There are 2 weeks that the machine is under l phases of repairing and the mean time of repairing in each phase is assumed to be equal, i.e. $\frac{1}{\mu_1} = \cdots = \frac{1}{\mu_l} = 2/l$. The mean time for arrival of demand is 1 week and the mean time for the machine to produce 1 unit of product is 1 day. Therefore we have $a = 1$ and $d = 7$. For simplicity, we set $N = m + h$. The probability density function for the demand is assumed to be geometric i.e. $q_i = (1-p)p^{i-1}$, $i = 1, 2, \cdots$. All the computations are done in a HP 712/80 workstation with MATLAB. We give the number of iterations for convergence of BGS (Tables 1 and 2) for different values of l and p. The numerical results indicated that the convergence rate for the BGS is independent of the numbers of phases l and increase very slowly with the parameters $N = m + h$.

Numerical Results of BGS for Strategy I & II

	$p = 0.5$								$p = 0.6$							
	$l=4$		$l=8$		$l=16$		$l=32$		$l=4$		$l=8$		$l=16$		$l=32$	
N	I	II	I	II	I	II	I	II	I	II	I	II	I	II	I	II
16	7	7	7	7	7	7	7	7	8	7	8	7	8	7	8	7
32	7	7	7	7	7	7	7	7	8	8	8	8	8	8	8	8
64	7	7	7	7	7	7	7	7	8	8	8	8	8	8	8	8
128	7	7	7	7	7	7	7	7	8	8	8	8	8	8	8	8
256	7	7	7	7	7	7	7	7	8	8	8	8	8	8	8	8
512	7	7	7	7	7	7	7	7	8	8	8	8	8	8	8	8

Table 1: Number of Iterations.

	$p = 0.7$								$p = 0.8$							
	$l=4$		$l=8$		$l=16$		$l=32$		$l=4$		$l=8$		$l=16$		$l=32$	
N	I	II	I	II	I	II	I	II	I	II	I	II	I	II	I	II
16	8	8	8	8	8	8	8	8	8	7	8	7	8	7	8	7
32	10	10	10	10	10	10	10	10	12	11	12	11	12	11	12	11
64	10	10	10	10	10	10	10	10	18	18	18	18	18	18	18	18
128	10	10	10	10	10	10	10	10	22	22	22	22	22	22	22	22
256	10	10	10	10	10	10	10	10	23	23	23	23	23	23	23	23
512	10	10	10	10	10	10	10	10	23	23	23	23	23	23	23	23

Table 2: Number of Iterations.

Next we consider the optimal hedging point h^* of the example, in which we keep the values of the parameters λ, μ_i, d and the assumptions for the demand as above, except that we set $l = 32$, $m = 50$ and the maximum inventory level to be 200. Moreover, the inventory cost c_I and backlog cost c_B

per unit of product are 50 and 2000, respectively. The backlog penalty cost c_P is 20000. For the demand parameter p, we test for some $p < p^* = \frac{298}{350}$, such that the average production rate is bigger than the mean demand rate. In the following tables (3 and 4), we give the optimal values of hedging point h^* and its corresponding average running cost per week for both Strategy I and Strategy II. We found that for the two strategies, the optimal hedging point h^* and the optimal average running cost are approximately the same in this example.

p	0.5	0.6	0.7	0.8
h^*	8	12	21	63
$C_1(h^*)$	485	703	1211	3406

Table 3: Optimal h^* and Expected Cost for Strategy I.

p	0.5	0.6	0.7	0.8
h^*	8	12	21	62
$C_2(h^*)$	485	703	1211	3331

Table 4: Optimal h^* and Expected Cost for Strategy II.

6 Concluding Remarks

A failure prone one-machine manufacturing system is considered in this paper. The machine-inventory process is modeled as an irreducible Markov chain. An efficient iterative method, the BGS method, is presented and analyzed to solve the steady state probability distribution of the process. Numerical experiments are reported to illustrate the convergence rate of the algorithm and the derivation of the optimal hedging points.

Two possible generalizations of our model are as follows.

- The maximum allowable backlog m is also considered as a decision variable for the optimization problem.

- The machine failure rate λ depends on the production rate. Note that in this case it has been shown that the optimal policy is still of hedging point type if λ is a linear function of the production rate, see Hu [12].

It would be interesting to extend our method to solve the above cases.

Appendix: The Algorithms

Let $\mathbf{x_0}$ be an initial guess of the solution of the non-singular system $B\mathbf{x} = \mathbf{b}$, then the CGS algorithm reads:

Conjugate Gradient Squared (CGS) Algorithm

$\mathbf{x} = \mathbf{x_0}$;
$\mathbf{r} = \mathbf{b} - B\mathbf{x}$;
$\tilde{\mathbf{r}} = \mathbf{s} = \mathbf{p} = \mathbf{r}$;
$\mathbf{w} = B\mathbf{p}$;
$\mu = \tilde{\mathbf{r}}^t \mathbf{r}$;
repeat the following :
  $\gamma = \mu$; $\quad \alpha = \gamma/\tilde{\mathbf{r}}^t \mathbf{r}$;
  $\mathbf{q} = \mathbf{s} - \alpha\mathbf{w}$; $\quad \mathbf{d} = \mathbf{s} + \mathbf{q}$;
  $\mathbf{w} = B\mathbf{d}$; $\quad \mathbf{x} = \mathbf{x} + \alpha\mathbf{d}$;
  $\mathbf{r} = \mathbf{r} - \alpha\mathbf{w}$;
  if converge then exit
  $\mu = \tilde{\mathbf{r}}^t \mathbf{r}$; $\quad \beta = \mu/\gamma$;
  $\mathbf{s} = \mathbf{r} - \beta\mathbf{q}$; $\quad \mathbf{p} = \mathbf{s} + \beta(\mathbf{q} + \beta\mathbf{p})$;
  $\mathbf{w} = B\mathbf{p}$;
end

Block Gauss-Seidel (BGS) Algorithm

$\mathbf{p}^0 = (p_1^0, \cdots, p_{l+1}^0)^t$ is an initial;
$\mathbf{p}^n = (p_1^n, \cdots, p_{l+1}^n)^t = $ the nth approximated solution.
$\mathbf{r}_n = $ the residual at the nth iteration.
$\mathbf{b} = (0, \cdots, 0, \delta)^t$;
$i = 0$;
repeat the following :
  $\mathbf{p}^{i+1} = (\lambda p_{l+1}^i, 0, \cdots, 0)^t + \mathbf{b}$;
  $p_1^{i+1} = (T_{10} + \mu_1 I)^{-1} p_1^{i+1}$;
  $for\ k = 2\ to\ l,$
  $p_k^{i+1} = (T_{10} + \mu_k I)^{-1}(p_k^{i+1} + \mu_{k-1} p_{k-1}^{i+1})$;
  end;
  $p_{l+1}^{i+1} = (T_{11} + \lambda I)^{-1}(p_l^{i+1} + \mu_l p_{k-1}^{i+1})$;
  if $\frac{\|r_{i+1}\|_2}{\|r_0\|_2} < $ tol then stop.
  $i = i + 1$;
end

References

[1] R. Akella and P. Kumar, (1986) *Optimal Control of Production Rate in a Failure Prone Manufacturing Systems*, IEEE Trans. Automat. Contr., 31, pp. 116–126.

[2] T. Bielecki and P. Kumar, (1988) *Optimality of Zero-inventory Policies for Unreliable Manufacturing Systems*, Operat. Res., 36, pp. 532–541.

[3] T. Chan, (1988) *An Optimal Circulant Preconditioner for Toeplitz Systems*, SIAM J. Sci. Statist. Comput., 9, pp. 767–771.

[4] R. Chan, (1987) *Iterative Methods for Overflow Queueing Models I*, Numer. Math. 51, pp. 143–180.

[5] R. Chan and W. Ching, *Toeplitz-circulant Preconditioners for Toeplitz Matrix and Its Application in Queueing Networks with Batch Arrivals*, To Appear in SAIM J. Sci. Statist. Comput..

[6] R. Chan and M. Yeung, (1993) *Circulant Preconditioners for Complex Toeplitz Matrices*, SIAM J. Numer. Anal. Vol. 30, No. 4, pp. 1193–1207.

[7] P. Davis, (1979) *Circulant Matrices*, John Wiley and Sons, New York.

[8] G. Golub and C. van Loan, (1983) *Matrix Computations*, John Hopkins University Press, Baltimore, MD.

[9] R. Horn and C. Johnson, (1991) *Matrix Analysis*, Cambridge University Press, Sydney.

[10] J. Hu and D. Xiang, (1993) *A Queueing Equivalence to Optimal Control of a Manufacturing System with Failures*, IEEE Trans. Automat. Contr., 38, pp. 499–502.

[11] J. Hu, *Production Rate Control for Failure Prone Production Systems with No Backlog Permitted*, To appear in IEEE Trans. on Automat. Contr.

[12] J. Hu, P. Vakili and G. Yu, (1991) *Optimality of Hedging Point Policies in the Production Control of Failure Prone Manufacturing Systems*, Submitted to IEEE Trans. on Automat. Contr.

[13] J. Hu and D. Xiang, (1991) *Structure Properties of Optimum Production Controllers of Failure Prone Manufacturing Systems*, Submitted.

[14] L. Kaufman, J. Seery and J. Morrison, (1981) *Overflow Models for Dimension PBX Feature Packages*, The Bell System Technical Journal, Vol. 60 No. 5.

[15] L. Kaufman, B. Gopinath and E. Wunderlich, (1981) *Analysis of Package Network Congestion Control Using Sparse Matrix Algorithms*, IEEE Trans. on Commun., Vol. 29, N0. 4, pp. 453–465.

[16] T. Ku and C. Kuo, (1991) *Design and Analysis of Toeplitz Preconditioners*, IEEE Trans. Signal Processing, 40, pp. 129–141.

[17] K. Kontovailis, R. Plemmons and W. Stewart, (1991) *Block Cyclic SOR for Markov Chains with p-Cyclic Infinitesimal Generator*, Lin. Alg. Appl., pp. 145–223.

[18] T. Oda, (1991) *Moment Analysis for Traffic Associated with Markovian Queueing Systems*, IEEE Trans. on Commun., 30, pp. 737–745.

[19] M. Posner and M. Berg, (1989) *Analysis of Production-inventory System with Unreliable Production Facility*, Operat. Res. Letters 8, pp. 339–345.

[20] C. Samaratunga, S. Sethi and X.Y. Zhou, *Computational Evaluation of Hierarchical Production Control Policies for Stochastic Manufacturing Systems*, To appear in Operat. Res.

[21] A. Seila, *Multivariate Estimation of Conditional Performance Measure in Regenerative Simulation*, (1990) American Journal of Mathematical and Management Sciences, 10, pp 17–45.

[22] S. Sethi, H. Yan, Q. Zhang and X.Y. Zhou, (1993) *Feedback Production Planning in a Stochastic Two-machine Flowshop: Asymptotic Analysis and Computational Results*, Int. J. Production Econ., 30-31, pp. 79-93.

[23] S. Sethi, Q. Zhang and X.Y. Zhou, (1992), *Hierarchical Controls in Stochastic Manufacturing Systems with Machines in Tandem*, Stoch. Stoch. Rep., 41, pp. 89-118.

[24] S. Sethi and X.Y. Zhou, (1994) *Dynamic Stochastic Job Shops and Hierarchical Production Planning*, IEEE Trans. Aut. Contr., AC-39, pp. 2061-2076.

[25] P. Sonneveld, (1989) *A Fast Lanczos-type Solver for Non-symmetric Linear Systems*, SIAM J. Sci. Statist. Comput., 10, pp. 36–52.

[26] G. Strang, (1986) *A Proposal for Toeplitz Matrix Calculations*, Stud. Appl. Math., 74, pp. 171–176.

[27] R. Varga, (1963) *Matrix Iterative Analysis*, Prentice-Hall, New Jersey.

[28] J. Wilkinson, (1965) *The Algebraic Eigenvalue Problem*, Clarendon Press, Oxford.

[29] H. Yan, X.Y. Zhou and G. Yin, (1994) *Finding Optimal Number of Kanbans in a Manufacturing System Via Stochastic Approximation and Perturbation Analysis*, Lecture Notes in Control and Information Sciences, Eds. G. Cohen and J.-P. Quadrat, Springer-Verlag, London, pp. 572-578.

Wai Ki Ching and Xun Yun Zhou
Department of Systems Engineering and
Engineering Management
The Chinese University of Hong Kong
Shatin, Hong Kong

Optimal Control of a Failure Prone Manufacturing System with the Possibility of Temporary Increase in Production Capacity*

Lei Huang   Jian-Qiang Hu   Pirooz Vakili

Abstract

Manufacturers have two general strategies to guard against uncertainties in the production or demand process: (a) building inventory to hedge against periods in which the production capacity is not sufficient to satisfy demand, and (b) temporarily increasing the production capacity, for example, by hiring temporary workers, adding overtime and/or extra shifts, or outsourcing part of the production. A production control problem is to strike a balance between the costs of building inventory, "buying" extra capacity, and backlogged demand. In this paper, we address this problem in the context of a single failure prone machine that produces a single part-type and show that the optimal control policy is a switching policy that is generally characterized by three switching levels.

1 Introduction

Uncertainty is one of the key ingredients that renders the planning and scheduling of manufacturing systems very difficult and highly challenging. From an aggregate point of view uncertainty manifests itself in two main areas: demand for products (volume, mix, and arrival times) and capacity of production (availability of raw material, operators, and machines). Manufacturers have two general strategies to guard against these uncertain elements in the demand and production processes: (a) building inventory to hedge against periods in which the production capacity is not sufficient to satisfy demand, and (b) temporarily increasing the production capacity, for example, by hiring

---

*Research was supported in part by the National Science Foundation under grants EID-9212122 and DDM-9215368.

temporary workers, adding overtime and/or extra shifts, or outsourcing part of the production. Clearly, each of these options has an associated cost and the task of a production planner and scheduler is to strike a balance between these costs, namely, the cost of building inventory, "buying" extra capacity, and unsatisfied demand. In this paper we address and solve this problem in the context of a simplified example.

We formulate the problem in the framework of flow control of manufacturing systems. In this framework, discrete parts are approximated by fluids and the movement of parts across the manufacturing floor is approximated by the flow of fluids among a network of capacitated machines. The changes in the demand and the production capacity are modeled by a discrete space stochastic process, often a Markov chain. The focus of most work reported in the literature has been the selection of the optimal production control policy when building inventory is the sole strategy to guard against uncertainty, i.e., it is assumed that the option of buying extra capacity is not available. In what follows we briefly review previous work on flow control of manufacturing systems; for a more detailed review and a comprehensive list of references on flow control of manufacturing systems, see two recent books by Gershwin [5] and Sethi and Zhang [22].

Olsder and Suri [19] were among the first to study the flow control problem for manufacturing systems and they proposed to use models with jump Markov disturbances based on Rishel's formalism ([20]). Kimemia and Gershwin [16] showed that the optimal control for such systems has a special structure called the hedging point policy. In such a policy, a nonnegative production surplus of part-types is maintained during times when excess capacity is available to hedge against future capacity shortages. The value of the optimal hedging point for systems with one machine and one part-type was first obtained by Akella and Kumar for the discounted cost problem [1] and by Bielecki and Kumar [2] for the average cost problem.

Sharifnia [23] showed how the optimum hedging points in the case of one part-type multiple machine states can be calculated. The problem of the complete evaluation of the optimal production policy in multiple part-types multiple machines is difficult because it requires solving either systems of partial differential equations or large dynamic programming problems that easily run into the "curse of dimensionality." Therefore, several approximation procedures have been proposed to obtain near-optimal controllers (e.g., [7, 3, 4]).

In all the work mentioned above, it has been assumed that the machine failure rates are independent of the production rates and are constant as long as the system is in one of its discrete capacity states, or in other words, the underlying Markov chain is homogeneous. Recently, several researchers [21, 10, 12, 13, 14, 18, 17, 24] have studied the problem of flow control for systems whose underlying state processes are not homogeneous Markov chains and established various structural properties for the optimal control policy. Hu and Xiang [11, 13, 8, 9] showed how some of these structural properties can be used to obtain the optimal control policy based on some known results in the area of stochastic process and queueing theory.

In this paper, we consider a problem in which it is possible to temporarily increase the production capacity at some additional cost. Being able to buy extra capacity is the main difference between our problem and those previously studied in the literature. This problem was first proposed by Gershwin [6]. A similar, but much simplified, problem was also considered by Hu [8] where it is assumed that backlog is not permitted, i.e., extra capacity has to be bought whenever the inventory level becomes zero. We will completely solve the problem in this paper. We prove that the optimal control policy is a policy with three thresholds: one corresponds to an inventory hedging level and the other two are levels below which extra production capacity should be bought to prevent further demand shortages (one for each machine state: up or down). Also, the optimal thresholds are obtained. It is interesting to notice that the optimal control policy is very much similar to the (S, s) inventory policy studied in inventory theory. In the (S, s) policy whenever the inventory level is below s, an order has to be placed to bring the inventory level back to S; in our problem, extra production capacity needs to be bought when the inventory level hits certain thresholds.

The rest of this paper is organized as follows. The problem formulation and some preliminaries are given in Section 2. In Section 3, a class of hedging point policies is defined and some of its properties are derived. The main results of the paper are given in Section 4. We conclude in Section 5.

2 Problem formulation and preliminaries

To put the problem we address in this paper in perspective, we begin with the description of a general version of it. Consider a production system that produces a number of product-types. Let $\mathbf{x} = (x_1, \cdots, x_n)$ denote the vector of inventory levels or backlogged demands for product-types (x_i is a continuous variable that represents the inventory level of product-type i if $x_i > 0$, or the backlogged demand for this product-type if $x_i < 0$). Let $\alpha = \{\alpha(t); t \geq 0\}$ denote the stochastic process representing random changes in the production capacity and random fluctuations in the demand rates for product-types (in general, α is a vector process). $(\mathbf{x}, \alpha) = \{(\mathbf{x}(t), \alpha(t)); t \geq 0\}$ is the state process of this system.

The production capacity at time t, denoted by $C_\alpha(t)$, is a random set and the vector of demand rates, denoted by $\mathbf{d}_\alpha(t)$ is a random vector. The control variables of production rates and rates of buying extra capacity for product-types at time t are denoted, respectively, by $\mathbf{u}(\mathbf{x}, \alpha)(t)$ and $\mathbf{v}(\mathbf{x}, \alpha)(t)$. The vector of feasible production rates at time t belongs to the random set $C_\alpha(t)$ and we assume that the vector of the rates of buying extra capacity belongs to a fixed set C. The dynamics of $\mathbf{x}$, the inventory-backlog vector, is given by

$$\frac{d\mathbf{x}}{dt}(t) = \mathbf{u}(\mathbf{x}, \alpha)(t) + \mathbf{v}(\mathbf{x}, \alpha)(t) - \mathbf{d}_\alpha(t). \tag{1}$$

Let $g(\mathbf{x})$ be the cost per unit time of holding $\mathbf{x}$ units of inventory or backlog, and let $h(\mathbf{v})$ be the cost per unit time of buying extra capacity at rate $\mathbf{v}$. Let

π belong to a feasible set of appropriately defined control policies Π, then we define the cost associated with policy π to be the long-run average cost given by

$$J^\pi = \lim_{T \to \infty} \frac{1}{T} E[\int_0^T (g(\mathbf{x}) + h(\mathbf{v}))dt \mid \mathbf{x}(0), \alpha(0)]. \qquad (2)$$

An optimal control policy in this setting is a feasible control policy $\pi^* \in \Pi$ that yields the minimum cost, i.e., such that

$$J^{\pi^*} \leq J^\pi \quad \text{for all } \pi \in \Pi. \qquad (3)$$

Having defined a general version of the problem, we now turn to specifying the assumptions that define the problem we address in this paper:

- $n = 1$, i.e., a single product-type is produced and x is a scalar.

- α is a time-homogeneous Markov chain on the set $\{1,2\}$. (We assume that the sojourn time of α in state i is exponentially distributed with rate q_i.) We assume that $\alpha = 2$ represents a state with "low" production capacity and $\alpha = 1$ represents a state with "high" production capacity; examples include the case where the production capacity C_α is deterministic and remains unchanged and the demand for the product alternates between periods of high demand ($\alpha = 2$) and low demand ($\alpha = 1$) according to a Markov chain α, or the case where the demand rate d_α is fixed and the production capacity alternates between periods of high capacity ($\alpha = 1$) and low capacity ($\alpha = 2$). To simplify the presentation and highlight the connection between the problem we consider here and others considered in the literature, we assume that the demand is fixed ($d_\alpha = d$) and that $\alpha = 2$ corresponds to the case where production capacity is equal to zero (down state). It is worth pointing out that solving this problem is equivalent to solving the others mentioned above.

- $C_1 = [0, r_1]$; we assume that $r_1 \geq d$, in other words, during the "up" periods the maximum possible production rate is higher than the demand rate;

$C_2 = \{0\}$;

$C = [0, r_2]$; we assume that $r_2 \geq d$.

- $g(x) = c^+ x^+ + c^- x^-$ where $x^+ = \max(x, 0)$, $x^- = \max(-x, 0)$, i.e., the inventory/backlog cost is piecewise linear;

  $h(v) = c_v v$, i.e., the cost of purchasing extra capacity is linear in the rate of purchase.

We limit ourselves to the class of stationary feedback control policies, where $\pi(x, \alpha, t) = \pi(x, \alpha) = (u(x, \alpha), v(x, \alpha))$.

Define the differential cost function $V_i^\pi(x)$ (value functions for simplicity) by

$$V_i^\pi(x) = \lim_{T \to \infty} \frac{1}{T} E[\int_0^T [(g(\mathbf{x}) + h(\mathbf{v}) - J^\pi])dt \mid \mathbf{x}(0) = x, \alpha(0) = i] \quad (4)$$

We use $V_i^*(x)$ ($i = 1, 2$) to denote optimal value functions. If we assume that they are differentiable, then the well-known Hamilton-Jacobi-Bellman equation for the optimal control is given by

$$\min_{0 \leq u \leq r_1,\ 0 \leq v \leq r_2} \{(u + v - d)\frac{dV_1^*(x)}{dx} + c_v v\} q_1(V_1^*(x) - V_2^*(x)) - g(x) + J^{\pi^*}$$

$$\min_{0 \leq v \leq r_2} \{(v - d)\frac{dV_2^*(x)}{dx} + c_v v\} q_2(V_1^*(x) - V_2^*(x)) - g(x) + J^{\pi^*}; \quad (5)$$

The following theorem tells us that the Hamilton-Jacobi-Bellman equation is a sufficient condition for a control to be optimal

Theorem 1 *If $(u, v) = \pi^*(x, \alpha)$, $V_1^{\pi^*}(x)$ and $V_2^{\pi^*}(x)$ satisfy Equation (5), then π^* is an optimal control policy.*

The proof of Theorem 1 is essentially the same as the one given in [2] which uses the Dynkin formula. Therefore, we shall not repeat it here.

3 Hedging point policies

In this section, we define a set of policies called *hedging point policies*. We will show subsequently that the optimal control policy belongs to this class of policies.

Each hedging point policy π^z is characterized by a vector of three parameters $z = (z_1, z_2, z_3)$ where $z_3 < z_2 \leq z_1$, and is defined as follows:

$$\pi^z(x,1) = (u,v) = \begin{cases} (0,0) & x > z_1 \\ (d,0) & x = z_1 \\ (r_1,0) & z_3 < x < z_1 \\ (r_1,r_2) & x \leq z_3, \end{cases}$$

$$\pi^z(x,0) = (u,v) = \begin{cases} (0,0) & x > z_2 \\ (0,d) & x = z_2 \\ (0,r_2) & x < z_2. \end{cases} \quad (6)$$

Note that, under the above policy, once the inventory/backlog is within the interval $[z_2, z_1]$, it remains in this interval for ever. In other words, all inventory/backlog levels in the intervals $(-\infty, z_2)$ and (z_1, ∞) are transient and are reached only a finite number of times. Thus J^z, the cost associated with policy π^z, is independent of z_3 and is only a function of z_1 and z_2. To make this fact explicit, we occasionally write $J(z_1, z_2)$ for J^z.

The following lemma implies that in search for the optimal hedging point policy among the class of hedging point policies, we only need to consider those policies for which $z_1 \geq 0$ and $z_2 \leq 0$.

Lemma 1 *If $0 < z_2 \leq z_1$, then $J(z_1 - z_2, 0) < J(z_1, z_2)$; if $z_2 \leq z_1 < 0$ then $J(0, z_2 - z_1) < J(z_1, z_2)$.*

Proof. Let π and π' be the hedging point policies defined by (z_1, z_2), where $0 < z_2 \leq z_1$, and by $(z_1 - z_2, 0)$, respectively (choose the third parameter of the hedging point policies arbitrarily). Let $X^\pi(0) = z_1$ and $X^{\pi'}(0) = z_1 - z_2$ and assume that the stochastic processes of the capacities are coupled, in the sense that the up and down periods of both processes coincide. Under these assumptions it is easy to verify that

$$X^{\pi'}(t) = X^\pi(t) - z_2.$$

Clearly there is no cost of backlog for either policy; moreover, under the above assumptions the durations of time when the policies purchase extra capacity and the corresponding rates of the purchase coincide and hence the associated costs are identical. However, in light of the above identity, the cost of inventory for policy π' is lower than that of policy π; hence $J^{\pi'} = J(z_1 - z_2, 0) < J^\pi = J(z_1, z_2)$.

A similar argument can be used to prove the result for the case where $z_2 \leq z_1 < 0$: Let π' be defined by $(0, z_2 - z_1)$, assume $X^\pi(0) = z_2$ and $X^{\pi'}(0) = z_2 - z_1$, and also couple the processes of the capacities as above. Then

$$X^{\pi'}(t) = X^\pi(t) - z_1.$$

In this case, there are no inventory costs for either policy, the costs of purchasing extra capacity are identical, and the cost of backlog is lower for policy π', hence $J^{\pi'} = J(0, z_2 - z_1) \leq J^\pi = J(z_1, z_2)$. □

In light of the above lemma, from now on, we limit ourselves to the class of hedging point policies for which $z_1 \geq 0 \geq z_2$. We denote this class by $\mathcal{H}$.

Our main results, given in the next section, are based on explicit evaluation of the average cost and the differential cost functions associated with a number of candidate hedging point policies. In preparation for what follows, we derive the average cost of an arbitrary policy $\pi^z \in \mathcal{H}$ and a set of differential equations that characterize the differential cost functions associated with. We begin with evaluating the average cost.

3.1 Average cost

Let $\pi^z \in \mathcal{H}$ be a hedging point policy. We define the following notation:

$\beta_1 = \frac{q_1}{r_1 - d}$,
$\beta_2 = \frac{q_2}{d}$,
$\lambda_0 = \beta_1 - \beta_2$,
$\rho = \frac{\beta_2}{\beta_1}$,
$p_1 = \frac{q_2}{q_1 + q_2}$,
$p_2 = \frac{q_1}{q_1 + q_2}$.

Let $\{X^z(t); t \geq 0\}$ be the process of inventory/backlog of the system, then we have

Proposition 1 $X^z(t)$ *converges weakly to the random variable of the steady state inventory/backlog denoted by* $X^z(\infty)$ *with probability density*

$$f(x) = \frac{r_1}{d}p_1\frac{\beta_1}{\beta_1 - \beta_2 e^{-\lambda_0(z_1-z_2)}}\lambda_0 e^{-\lambda_0(x-z_2)} \quad z_2 < x < z_1,$$

$$P_{z_1} = P(X^z(\infty) = z_1) = p_1\frac{1}{\beta_1 - \beta_2 e^{-\lambda_0(z_1-z_2)}}\lambda_0 e^{-\lambda_0(z_1-z_2)},$$

$$P_{z_2} = P(X^z(\infty) = z_2) = p_2\frac{1}{\beta_1 - \beta_2 e^{-\lambda_0(z_1-z_2)}}\lambda_0.$$

Proof. The same results are obtained in [8] for the case $z_2 = 0$. In [8] it is shown that $X^z(t)$ is related to the workload process of an M/M/1 queue with limited workload, based on which the density function of $X^z(t)$ can then be obtained by using the existing results in queueing theory (also see [11]). For $z_2 < 0$, we can apply the results in [8] to the process $X^z(t) - z_2$, then shift everything by z_2. □

Given the above steady state density, it is straightforward to evaluate the average cost associated with policy π^z. Let J_1^z = cost of holding inventory, J_2^z = cost of having backlogged demand, and J_3^z = cost of purchasing extra capacity. Then

$$J_1^z = c^+ p_1 \frac{1}{\beta_1 - \beta_2 e^{-\lambda_0(z_1-z_2)}}\{(\lambda_0 - \frac{r_1}{d}\beta_1)z_1 e^{-\lambda_0(z_1-z_2)} - \frac{r_1}{d}\beta_1 \lambda_0 e^{-\lambda_0(z_1-z_2)}$$
$$+ \frac{r_1}{d}\beta_1 \frac{1}{\lambda_0} e^{\lambda_0 z_2}\},$$

$$J_2^z = c^- \frac{1}{\beta_1 - \beta_2 e^{-\lambda_0(z_1-z_2)}}\{(\frac{r_1}{d}\beta_1 p_1 + p_2 \lambda_0)z_2 - \frac{r_1}{d}\beta_1 p_1 \frac{1}{\lambda_0} e^{\lambda_0 z_2} + \frac{r_1}{d}\beta_1 p_1 \frac{1}{\lambda_0}\},$$

$$J_3^z = c_v d p_2 \frac{\lambda_0}{\beta_1 - \beta_2 e^{-\lambda_0(z_1-z_2)}}.$$

The total cost, $J^z = J(z_1, z_2) = J_1^z + J_2^z + J_3^z$. It is worth noting that $J(z_1, z_2)$ is, clearly, a continuously differentiable function of z_1 and z_2 on the region $\{z_1 \geq 0, z_2 \leq 0\}$.

3.2 Value function

Let $V^z(x) = (V_1^z(x), V_2^z(x))'$ be the vector of value functions corresponding to policy π^z ($\prime$ denotes transpose). Also let

$$A_1 = \begin{bmatrix} -\frac{q_1}{d} & \frac{q_1}{d} \\ \frac{q_2}{d} & -\frac{q_2}{d} \end{bmatrix} \quad b_1 = \begin{bmatrix} \frac{1}{d} \\ \frac{1}{d} \end{bmatrix} \quad A_0 = \begin{bmatrix} \frac{q_1}{r_1-d} & -\frac{q_1}{r_1-d} \\ \frac{q_2}{d} & -\frac{q_2}{d} \end{bmatrix} \quad b_0 = \begin{bmatrix} -\frac{1}{r_1-d} \\ \frac{1}{d} \end{bmatrix}$$

$$A_2 = \begin{bmatrix} \frac{q_1}{r_1-d} & -\frac{q_1}{r_1-d} \\ -\frac{q_2}{r_2-d} & \frac{q_2}{r_2-d} \end{bmatrix} \quad b_2 = \begin{bmatrix} -\frac{1}{r_1-d} \\ -\frac{1}{r_2-d} \end{bmatrix} \quad b'_2 = \begin{bmatrix} 0 \\ -\frac{1}{r_2-d} \end{bmatrix}$$

$$A_3 = \begin{bmatrix} \frac{q_1}{r_2+r_1-d} & -\frac{q_1}{r_2+r_1-d} \\ -\frac{q_2}{r_2-d} & \frac{q_2}{r_2-d} \end{bmatrix} \quad b_3 = \begin{bmatrix} -\frac{1}{r_1+r_2-d} \\ -\frac{1}{r_2-d} \end{bmatrix}.$$

We have

Proposition 2 *The value function V^z satisfies the following set of piecewise linear differential equations:*

$$\begin{aligned}
\frac{dV^z(x)}{dx} &= A_1 V^z(x) - b_1(J^z - c^+ x); & z_1 &\leq x \\
\frac{dV^z(x)}{dx} &= A_0 V^z(x) - b_0(J^z - c^+ x); & 0 &\leq x < z_1 \\
\frac{dV^z(x)}{dx} &= A_0 V^z(x) - b_0(J^z + c^- x); & z_2 &\leq x < 0 \\
\frac{dV^z(x)}{dx} &= A_2 V^z(x) - b_2(J^z + c^- x) + b'_2 c_v r_2; & z_3 &\leq x < z_2 \\
\frac{dV^z(x)}{dx} &= A_3 V^z(x) - b_3(J^z + c^- x) + b_3 c_v r_2; & x &< z_3
\end{aligned} \qquad (7)$$

with the following boundary conditions:

$$\begin{aligned}
V_1^z(z_1) - V_2^z(z_1) &= \frac{c^+ z_1 - J^z}{q_1} \\
V_1^z(z_2) - V_2^z(z_2) &= \frac{c^- z_2 + J^z - c_v d}{q_2}
\end{aligned} \qquad (8)$$

Proof. It is very straightforward, e.g., see [2].

4 Optimal Control Policies

In this section we give the main results of this paper and show that the optimal control policies belong to the class of hedging point policies. Four cases are identified that correspond to: (1) the hedging levels for inventory and backlog are both non-zero, (2) the hedging level for inventory is zero but the backlog hedging level is nonzero, (3) the hedging level for backlog is zero but the inventory hedging level is nonzero, and, finally, (4) both hedging levels are zero.

In each case we first identify necessary conditions for the existence of a hedging point policy that is optimal among an appropriate subset of hedging point policies. This yields a candidate policy. We then identify sufficient conditions under which the candidate policy is optimal among all control policies.

4.1 Case 1: Nonzero inventory, nonzero backlog ($z_1^* > 0$, $z_2^* < 0$)

A necessary condition for the existence of a hedging point policy π^{z^*}, with $z_1^* > 0$, $z_2^* < 0$, that is optimal among all feasible policies is that the policy be optimal among all hedging point policies. Given the fact that the cost function $J(z_1, z_2)$ is differentiable, a set of necessary conditions for the latter to be true is the existence of two hedging levels $z_1^* > 0$, $z_2^* < 0$, such that

$$\frac{\partial J}{\partial z_1}(z_1^*) = 0, \quad \frac{\partial J}{\partial z_2}(z_2^*) = 0. \tag{9}$$

Taking derivatives, setting them equal to zero and rearranging terms yields the following set of equations that is equivalent to (9) in the sense that the existence of $z_1^* > 0$, $z_2^* < 0$ satisfying one set of equations implies that $z_1^* > 0$ and $z_2^* < 0$ satisfy the other,

$$\begin{aligned} f_1(z_1^*) &= c^+\rho e^{-\lambda_0 z_1^*} + c^- e^{-\frac{\lambda_0}{c^+}(k - c^+ z_1^*)} - \frac{r_1}{d}p_1(c^+ + c^-) = 0; \\ f_2(z_2^*) &= c^+\rho e^{-\frac{\lambda_0}{c^+}(k - c^- z_2^*)} + c^- e^{-\lambda_0 z_2^*} - \frac{r_1}{d}p_1(c^+ + c^-) = 0, \end{aligned} \tag{10}$$

where $k = dp_2[\rho(\frac{c^-}{q_2} - c_v) - (\frac{c^{\pm}}{q_1} - c_v)]$.

Proposition 3 *A set of necessary and sufficient conditions for the existence of a pair of values $z_1^* > 0$, $z_2^* < 0$ that satisfy equation(9) is $f_1(0) < 0$ and $f_2(0) < 0$. The vector (z_1^*, z_2^*) is unique when it exists.*

Proof. Note that f_1 and f_2 are linear combinations of convex functions with non-negative coefficients, therefore they are convex (both are of the form $a_1 e^{b_1 z} + a_2 e^{-b_2 z} + a_3$); moreover, $f_i(\infty) = f_i(-\infty) = \infty; i = 1, 2$ (note that $b_1, b_2, a_1, a_2 > 0$). The minimizers of the functions f_1 and f_2, denoted by z_1^m

and z_1^m respectively, are $z_1^m = \frac{c^- \ln \rho + \lambda_0 k}{\lambda_0 (c^+ + c^-)}$, and $z_2^m = \frac{c^+ \ln \rho - \lambda_0 k}{\lambda_0 (c^+ + c^-)}$; the minimums of these functions are respectively $f_1(z_1^m) = (c^+ + c^-)[\rho e^{-\lambda_0 z_1^m} - \frac{r_1}{d} p_1]$, $f_2(z_2^m) = (c^+ + c^-)[e^{-\lambda_0 z_2^m} - \frac{r_1}{d} p_1]$.

Let $f_1(0) < 0$ and $f_2(0) < 0$, then given the above properties of the two functions f_1 and f_2, it can be easily verified that f_1 has a unique root $z_1^* > 0$ and f_2 has a unique root $z_2^* < 0$.

On the other hand, assume that $z_1^* > 0$. If $z_1^m \leq 0$, then $f_1(z_1^m) < 0$ (recall $f_1(z_1^*) = 0$) and clearly $f_1(0) < 0$; Assume $z_1^m > 0$; further assume that $\rho > 1$, in this case $\lambda_0 < 0$ and $\lambda_0 z_1^m < 0$, and $\rho e^{-\lambda_0 z_1^m} > \rho$, therefore

$$f_1(z_1^m) = (c^+ + c^-)[\rho e^{-\lambda_0 z_1^m} - \frac{r_1}{d} p_1] > (c^+ + c^-)[\rho - \frac{r_1}{d} p_1] = (c^+ + c^-) p_1 (\rho - 1) > 0$$

which contradicts the existence of $z_1^* > 0$; hence $z_1^* > 0$ and $z_1^m > 0$, implies $\rho < 1$. $\rho < 1$ is equivalent to $\lambda_0 > 0$; this latter condition and $z_1^m > 0$ yield $c^- \ln \rho + \lambda_0 k > 0$, therefore $\ln \rho > -\frac{\lambda_0 k}{c^-}$; hence

$$f_1(0) = c^+ \rho + c^- e^{-\frac{\lambda_0 k}{c^-}} - \frac{r_1}{d} p_1 (c^+ + c^-)$$
$$< c^+ \rho + c^- \rho - \frac{r_1}{d} p_1 (c^+ + c^-) = (c^+ + c^-) p_1 (\rho - 1) < 0.$$

Now assume that $z_2^* < 0$. If $z_2^m \geq 0$, then $f_2(z_2^m) < 0$ (recall $f_2(z_2^*) = 0$) and clearly $f_2(0) < 0$; Assume $z_2^m < 0$; further assume that $\rho < 1$, in this case $\lambda_0 > 0$ and $-\lambda_0 z_1^m > 0$, and $\rho e^{-\lambda_0 z_1^m} > \rho$, therefore,

$$f_2(z_2^m) = (c^+ + c^-)[\rho e^{-\lambda_0 z_2^m} - \frac{r_1}{d} p_1]$$
$$> (c^+ + c^-)[\rho - \frac{r_1}{d} p_1] = (c^+ + c^-) p_1 (\rho - 1) > 0$$

which contradicts the existence of $z_2^* < 0$; hence $z_2^* < 0$ and $z_2^m < 0$, implies $\rho > 1$. $\rho > 1$ is equivalent to $\lambda_0 < 0$; this latter condition and $z_2^m < 0$ yield $-c^+ \ln \rho + \lambda_0 k > 0$, therefore $-\ln \rho > -\frac{\lambda_0 k}{c^+}$; hence

$$f_2(0) = c^+ \rho e^{-\frac{\lambda_0 k}{c^+}} + c^- - \frac{r_1}{d} p_1 (c^+ + c^-)$$
$$< c^+ + c^- - \frac{r_1}{d} p_1 (c^+ + c^-) = (c^+ + c^-) p_1 (1 - \rho) < 0.$$

□

The following corollary is one of the main results of this subsection:

Theorem 2 *A set of necessary conditions for the existence of a hedging point policy π^{z*} that is optimal among all feasible policies and such that $z_1^* > 0, z_2^* < 0$ is*

$$f_1(0) < 0, \quad f_2(0) < 0. \tag{11}$$

We now turn to deriving sufficient conditions for the existence of such a hedging point policy π^{z*}. We will show that the above conditions are also sufficient. In order to accomplish this, we first need to specify the policy π^{z*} completely. Note that while the set of conditions (9) uniquely defines z_1^* and z_2^*, it does not specify the hedging point policy uniquely. This is because the third hedging point z_3^* is undefined. To define z_3^*, we begin with an arbitrary $z_3 < z_2^*$ and consider the hedging point policy associated with (z_1^*, z_2^*, z_3). We denote the cost associated with this policy by J^*. Note that J^* is independent of z_3 and is uniquely defined by z_1^* and z_2^*. Using equations (9), J^* can be written in terms of either z_1^* or z_2^* as

$$\begin{aligned} J^* &= c^+ z_1^* + c^+ \frac{d}{q_1 + q_2}, \\ J^* &= c^-(-z_2^*) + c^- \frac{r_1 - d}{q_1 + q_2} + c_v(d - \frac{q_2 r_1}{q_1 + q_2}). \end{aligned} \tag{12}$$

To simplify the presentation and with a slight abuse of notation, we denote the value functions associated with the above hedging point policy by V_1^* and V_2^*.

Using the set of piecewise linear differential equations (7), the boundary conditions (8), and the above identities, we can determine $\frac{dV_1^*(x)}{dx}$ and $\frac{dV_2^*(x)}{dx}$. Before specifying these functions, we introduce the following notation:

$\lambda_1 = \frac{q_1}{d} + \frac{q_2}{d};$
$\lambda_0 = \frac{q_1}{r_1 - d} - \frac{q_2}{d};$
$\lambda_2 = \frac{q_1}{r_1 - d} + \frac{q_2}{r_2 - d};$
$\lambda_3 = \frac{q_1}{r_1 + r_2 - d} + \frac{q_2}{r_2 - d}.$

Let $R_1 = \{x; z_1^* \leq x\}$, $R_0 = \{x; z_2^* \leq x < z_1^*\}$, $R_2 = \{x; z_3 \leq x < z_2^*\}$, and $R_3 = \{x; x \leq z_3\}$. Then $\frac{1}{\lambda_i}$ = average change in the inventory/backlog per one up and down period in region R_i, assuming the trajectory remains in R_i during this time. Given this notation, we have

For $z_1^* \leq x$:

$$\frac{dV_1^*(x)}{dx} = \frac{c^+(q_1 + q_2)}{d^2}[\frac{1}{\lambda_1}(x - z_1^*) - \frac{1}{\lambda_1^2}(1 - e^{-\lambda_1(x-z_1^*)})];$$

$$\frac{dV_2^*(x)}{dx} = \frac{c^+(q_1 + q_2)}{d^2}[\frac{1}{\lambda_1}(x - z_1^*) + \frac{q_1}{q_2}\frac{1}{\lambda_1^2}(1 - e^{-\lambda_1(x-z_1^*)})] - \frac{c^+}{q_1}.$$

For $0 \leq x < z_1^*$:

$$\frac{dV_1^*(x)}{dx} = \frac{c^+(q_1 + q_2)}{d(r_1 - d)}[-\frac{1}{\lambda_0}(z_1^* - x) + \frac{1}{\lambda_0^2}(1 - e^{-\lambda_0(z_1^*-x)})];$$

$$\frac{dV_2^*(x)}{dx} = \frac{c^+(q_1 + q_2)}{d(r_1 - d)}[-\frac{1}{\lambda_0}(z_1^* - x) + \frac{\rho}{\lambda_0^2}(1 - e^{-\lambda_0(z_1^*-x)})] - \frac{c^+}{q_1}.$$

For $z_2^* \leq x < 0$:

$$\frac{dV_1^*(x)}{dx} = \frac{c^-(q_1 + q_2)}{d(r_1 - d)}[\frac{1}{\lambda_0}(x - z_2^*) - \frac{1}{\rho\lambda_0^2}(1 - e^{-\lambda_0(x-z_2^*)})] - c_v + \frac{c^-}{q_2};$$

$$\frac{dV_2^*(x)}{dx} = \frac{c^-(q_1 + q_2)}{d(r_1 - d)}[-\frac{1}{\lambda_0}(x - z_2^*)$$

$$-\frac{1}{\lambda_0^2}(1 - e^{-\lambda_0(x-z_2^*)})] - c_v.$$

For $z_3 \leq x < z_2^*$:

$$\frac{dV_1^*(x)}{dx} = \frac{c^-(q_1 + q_2)}{(r_1 - d)(r_2 - d)}[-\frac{1}{\lambda_2}(z_2^* - x) - \frac{q_1(r_2 - d)}{q_2(r_1 - d)}$$

$$\times \frac{1}{\lambda_2^2}(1 - e^{-\lambda_2(z_2^*-x)})] - c_v + \frac{c^-}{q_2};$$

$$\frac{dV_2^*(x)}{dx} = \frac{c^-(q_1 + q_2)}{(r_1 - d)(r_2 - d)}[-\frac{1}{\lambda_2}(z_2^* - x) - \frac{1}{\lambda_2^2}(1 - e^{-\lambda_2(z_2^*-x)})] - c_v.$$

We are now prepared to define z_3^*. Taking derivative of $\frac{dV_1^*(x)}{dx}$ over the interval $z_3 \leq x < z_2^*$, we have

$$\frac{d^2V_1^*(x)}{dx^2} = \frac{c^-(q_1 + q_2)}{d(r_1 - d)}[\frac{1}{\lambda_2} + \frac{q_1}{q_2}\frac{1}{\lambda_2}(1 - e^{-\lambda_2(z_2^*-x)})].$$

Given that $\lambda_2 = \frac{q_1}{r_1-d} + \frac{q_2}{r_2-d} > 0$, it follows that $\frac{d^2V_1^*(x)}{dx^2} > 0$ on the interval $[z_3, z_2^*)$; hence, on this interval, $\frac{dV_1^*(x)}{dx}$ is continuous and strictly decreasing.

Moreover, we have $\frac{dV_1^*(z_2^*)}{dx} = -c_v + \frac{c^-}{q_2}$ and $\lim_{x \to -\infty} \frac{dV_1^*(x)}{dx} = -\infty$ (since z_3 is arbitrary, assume $z_3 \to -\infty$ as $x \to -\infty$). Therefore, there exists a unique value at which the derivative of V_1^* is equal to $-c_v$. We define z_3^* to be this unique value, i.e., z_3^* is defined uniquely via the identity

$$\frac{dV_1^*(z_3^*)}{dx} = -c_v.$$

We have now completely specified the policy π^{z^*} which we claim is optimal among all feasible policies. We denote the value functions associated with this policy also by V_1^* and V_2^*. On the interval (z_3^*, ∞) the derivatives of these functions are the same as the ones calculated above for the hedging point policy with parameters (z_1^*, z_2^*, z_3), (replace z_3 by z_3^*). On the last region, R_3, the derivatives are given by:

For $x \leq z_3^*$:

$$\begin{aligned}
\frac{dV_1^*(x)}{dx} &= \frac{c^-(q_1+q_2)}{(r_1+r_2-d)(r_2-d)}[-\frac{1}{\lambda_3}(z_3^*-x) + \frac{r_1 q_1}{(q_1+q_2)(r_1+r_2-d)} \\
&\quad \times \frac{1}{\lambda_3^2}(1-e^{-\lambda_3(z_3^*-x)})] \\
&\quad + \frac{c^-(q_1+q_2)}{(r_1+r_2-d)(r_2-d)}[-\frac{1}{\lambda_3}(z_2^*-z_3^*) + \frac{r_1-d}{q_1+q_2}] \\
&\quad (1-e^{-\lambda_3(z_3^*-x)}) - c_v, \\
\frac{dV_2^*(x)}{dx} &= \frac{c^-(q_1+q_2)}{(r_1+r_2-d)(r_2-d)}[-\frac{1}{\lambda_3}(z_3^*-x) + \frac{r_1 q_2}{(q_1+q_2)(r_2-d)} \\
&\quad \times \frac{1}{\lambda_3^2}(1-e^{-\lambda_3(z_3^*-x)})] \\
&\quad + \frac{c^-(q_1+q_2)}{(r_2-d)^2}\frac{q_2}{q_1}[-\frac{1}{\lambda_3}(z_2^*-z_3^*) + \frac{r_1-d}{q_1+q_2}](1-e^{-\lambda_3(z_3^*-x)}) \\
&\quad - \frac{c^-(q_1+q_2)}{q_1(r_2-d)}[(z_2^*-z_3^*) - (r_1-d)] - c_v.
\end{aligned}$$

Proposition 4 *Let $V_i^*(i = 1, 2)$ be the value functions associated with the hedging point policy defined by z_1^*, z_2^* and z_3^* given above; then V_i^* is convex and continuously differentiable $(i = 1, 2)$.*

Proof. The continuity of the derivatives at all points except at $x = 0$ and $x = z_3^*$ is obvious; the continuity at $x = 0$ follows from the identities $f_1(z_1^*) = 0$ and $f_2(z_2^*) = 0$ while the continuity at $x = z_3^*$ can be easily verified by using $dV_1^*(z_3^*)/dx = -c_v$. The proof of convexity is straightforward for $x > z_3^*$. So we focus on the convexity for $x \leq z_3^*$. First for $x \leq z_3^*$ we have

$$\frac{d^2 V_1^*(x)}{dx^2} = \frac{c^-}{(r_1 + r_2 - d)(r_2 - d)}((q_1 + q_2)(z_2^* - z_3^*) + (r_2 - r_1))e^{-\lambda_3(z_3^* - x)}$$
$$+ \frac{c^-(q_1 + q_2)}{\lambda_3(r_1 + r_2 - d)(r_2 - d)}(1 - e^{-\lambda_3(z_3^* - x)})$$

$$\frac{d^2 V_2^*(x)}{dx^2} = \frac{c^-}{q_1(r_1 + r_2 - d)(r_2 - d)}(-q_2(q_1 + q_2)(z_2^* - z_3^*)$$
$$+ q_1(r_2 - d) + q_2(r_1 - d))e^{-\lambda_3(z_3^* - x)}$$
$$+ \frac{c^-(q_1 + q_2)}{\lambda_3(r_1 + r_2 - d)(r_2 - d)}(1 - e^{-\lambda_3(z_3^* - x)})$$

To prove $d^2 V_1^*(x)/dx^2 \geq 0$ and $d^2 V_2^*(x)/dx^2 \geq 0$, it is sufficient to show that $(q_1+q_2)(z_2^* - z_3^*) + (r_2 - r_1) \geq 0$ and $-q_2(q_1+q_2)(z_2^* - z_3^*) + q_1(r_2 - d) + q_2(r_1 - d) \geq 0$. To verify these two inequalities, we consider the interval $[z_3^*, z_2^*]$. Since

$$\frac{dV_1^*(z_2^*)}{dx} = \frac{c^-}{q_2} - c_v \quad \text{and} \quad \frac{dV_1^*(z_3^*)}{dx} = -c_v$$

we have

$$\frac{c^-}{q_2} = \int_{z_3^*}^{z_2^*} \frac{d^2 V_1^*(x)}{dx^2} dx$$
$$= \frac{c^- q_1(q_1 + q_2)}{\lambda_2^2 q_2 (r_1 - d)^2}(1 - e^{-\lambda_2(z_2^* - z_3^*)}) + \frac{c^-(q_1 + q_2)(z_2^* - z_3^*)}{q_1(r_2 - d) + q_2(r_1 - d)}. \quad (13)$$

Based on (13), we have

$$\frac{c^-}{q_2} = \frac{c^- q_1(q_1 + q_2)}{\lambda_2^2 q_2 (r_1 - d)^2}(1 - \lambda_2(z_2^* - z_3^*) - e^{-\lambda_2(z_2^* - z_3^*)}) + \frac{c^-(q_1 + q_2)(z_2^* - z_3^*)}{q_2(r_1 - d)}$$
$$\leq \frac{c^-(q_1 + q_2)(z_2^* - z_3^*)}{q_2(r_1 - d)},$$

hence, $(q_1 + q_2)(z_2^* - z_3^*) + (r_2 - r_1) \geq 0$. On the other hand, (13) also leads to

$$\frac{c^-}{q_2} \geq \frac{c^-(q_1 + q_2)(z_2^* - z_3^*)}{q_1(r_2 - d) + q_2(r_1 - d)},$$

which gives $-q_2(q_1 + q_2)(z_2^* - z_3^*) + q_1(r_2 - d) + q_2(r_1 - d) \geq 0$. This completes our proof. □

We are now prepared to state the other main result of this subsection:

Theorem 3 *The set of conditions (11) is sufficient for the existence of a hedging point policy with nonzero hedging levels that is optimal among all feasible policies.*

Proof. Assume that conditions (11) are satisfied. Then by proposition (3) there exist unique hedging levels $z_1^* > 0$ and $z_2^* < 0$ that satisfy equations (9). According to the preceding discussion, we can then define z_3^* uniquely. The proof of the theorem is complete if we show that the hedging point policy defined by (z_1^*, z_2^*, z_3^*) satisfies the Hamilton-Jacoby-Bellman (HJB) equations.

For $\alpha = 1$, i.e., when the machine is up, using formulas for the derivatives of the value functions, and Proposition (4), it is easy to verify that $\frac{dV_1^*(x)}{dx}$ switches sign from negative to positive at $x = z_1^*$, and $\frac{dV_1^*(x)}{dx} + c_v$ switches sign from negative to positive at z_3^*. Therefore, the policy $u(x,1) = r_1$, $x < z_1^*$ and $u(x,1) = 0$, $x > z_1^*$ and $v(x,1) = r_2$, $x \leq z_3^*$ and $v(x,1) = 0$, $x > z_3^*$ solves the following minimization problem:

$$\min_{0 \leq u \leq r_1,\ 0 \leq v \leq r_2} \{(u+v-d)\frac{dV_1^*(x)}{dx} + c_v v\}.$$

It is also easy to verify that $\frac{dV_1^*(x)}{dx} + c_v$ switches sign from negative to positive at z_2^*; hence the policy $v(x,2) = r_2$, $x < z_2^*$ and $v(x,2) = 0$, $x > z_2^*$ solves the following minimization problem:

$$\min_{0 \leq v \leq r_2} \{(v-d)\frac{dV_2^*(x)}{dx} + c_v v\}.$$

The minimizing policy is defined everywhere except at $u(z_1^*, 1)$ and at $v(z_2^*, 2)$. To avoid chattering, we set $u(z_1^*, 1) = v(z_2^*, 2) = d$. This policy is exactly the hedging point policy with parameters (z_1^*, z_2^*, z_3^*). The HJB equation is therefore satisfied by this hedging point policy. □

Before closing this subsection, we provide some discussion on the conditions $f_1(0) < 0$ and $f_2(0) < 0$. First, we present the following result:

Proposition 5 *The conditions $f_1(0) < 0$ and $f_2(0) < 0$ imply $c_v > c^+/q_1$ and $c_v > c^-/q_2$, respectively.*

Proof. We will only give the proof for $c_v > c^+/q_1$; the proof for $c_v > c^-/q_2$ is essentially the same. Using the inequality $1 + x \le e^x$, we have

$$c^+\rho + c^-(1 - \frac{\lambda_0 k}{c^-}) - \frac{r_1}{d}p_1(c^+ + c^-) \le f_1(0) < 0.$$

With some simplification, this leads to

$$\frac{(1-\rho)^2 dq_1}{(r_1 - d)(q_1 + q_2)}(c^+ - c_v q_1) < 0,$$

which implies $c_v > c^+/q_1$. $\square$

Based on the above proposition, it is clear that $c_v > c^+/q_1$ and $c_v > c^-/q_2$ are necessary conditions for the existence of a hedging point policy π^{z*} with $z_1^* > 0, z_2^* < 0$ that is optimal among all feasible policies. In the following subsections we will show that the relationships between c_v and c^+/q_1, c^-/q_2 also give necessary and sufficient conditions for the other three cases.

4.2 Zero inventory, nonzero backlog ($z_1^* = 0,\ z_2^* < 0$)

We now turn to the case where the optimal inventory hedging level is zero and that of backlog is strictly negative. A necessary condition for the existence of a hedging point policy with these properties is that it be optimal among all hedging point policies with zero inventory hedging levels; a necessary condition for the latter to be true is

$$\frac{\partial J(0, z_2)}{\partial z_2}\bigg|_{z_2 = z_2^*} = 0.$$

Taking derivative of $J(0, z_2)$ with respect to z_2, simplifying, and setting it equal to zero, yields the following equation:

$$f_3(z_2) = \frac{1}{\lambda_0}[c^-\lambda_0 z_2 + \frac{c^-}{\rho}(e^{-\lambda_0 z_2} - 1) + \frac{d(r_1 - d)}{q_1 + q_2}\lambda_0^2(\frac{c^-}{q_2} - c_v)] = 0.$$

Proposition 6 *A necessary and sufficient condition for the existence of a negative root for $f_3(z_2) = 0$ is*

$$c_v > \frac{c^-}{q_2};$$

such a negative root is unique.

Proof. Let $g_1(z_2) = \lambda_0 f_3(z_2)$. The roots of f_3 and g_1 coincide, so we can focus on the function g_1. Note that $\frac{d^2 g_1}{dz_2^2}(z_2) = \frac{c^-}{\rho}\lambda_0^2 e^{-\lambda_0 z_2} > 0$, hence g_1 is convex. Moreover, it is easy to verify that the minimizer of g_1 is at $z_2^m = -\frac{\ln \rho}{\lambda_0}$; note that $\rho < 1 \Leftrightarrow \lambda_0 < 0$, therefore $z_2^m > 0$. Hence, a necessary and sufficient condition for the existence of a negative root for g_1 is $g_1(0) = \frac{d(r_1-d)}{q_1+q_2}\lambda_0^2(\frac{c^-}{q_2} - c_v) < 0$, which is equivalent to $c_v > \frac{c^-}{q_2}$. Clearly, in this case, the negative root of g_1, and hence of f_3, is unique. □

We can now give one of the main results of this subsection which is a corollary of the above proposition.

Theorem 4 *A necessary condition for the existence of a hedging point policy π^{z*} with $z_1^* = 0, z_2^* < 0$, that is optimal among all feasible policies, is*

$$c_v > \frac{c^-}{q_2}. \tag{14}$$

Before looking at the value functions, we prove a result about the hedging level z_2^* which will be needed in what follows.

Lemma 2 *Assume $c_v > \frac{c^-}{q_2}$, and let z_2^* be as defined above. Then*

$$\frac{q_1+q_2}{q_2(r_1-d)}c^- z_2^* > \frac{c^-}{q_2} - c_v. \tag{15}$$

Proof. Assume $\lambda_0 > 0$. Using the inequality $(e^{-\lambda_0 x} - 1) > -\lambda_0 x$ in the equation $f_3(z_2^*) = 0$, we get

$$c^- z_2^* - \frac{1}{\rho}c^- z_2^* < -\frac{d(r_1-d)}{q_1+q_2}\lambda_0(\frac{c^-}{q_2} - c_v).$$

The result follows immediately from the above inequality. The case of $\lambda_0 < 0$ can be proved similarly. □

Let 0 and z_2^* denote the optimal hedging levels (when they exist) and let J^* denote the corresponding cost; J^* is given by

$$J^* = c^-(-z_2^*) + c^- \frac{r_1 - d}{q_1 + q_2} + c_v d\left(1 - \frac{q_2 r_1}{d(q_1 + q_2)}\right). \tag{16}$$

Note that the above cost function J^* and that for the case $z_1^* > 0, z_2^* < 0$ (equation (12), J^* expressed in terms of z_2^*) are identical in form (the optimal z_2^* for the two cases are in general different). Note also that the value functions are obtained from solving the set of differential equations (7) after substituting J^* for J^z (J^* expressed in terms of z_2^* is substituted on the interval $(-\infty, 0)$). Hence, given the similarity in form of J^* in the two cases, the value functions V_1^* and V_2^* are also similar in form on the interval $(-\infty, 0)$. As a result, following the same steps as in the previous section, we can define z_3^*, and the same arguments as those given in the previous section can be used to show that V_1^* and V_2^* are convex and continuously differentiable on the interval $(-\infty, 0)$. We now focus on the interval $[0, \infty)$.

For $0 \leq x$:

$$\frac{dV_1^*(x)}{dx} = \frac{c^+}{d}x - \frac{J^*}{d}(1 - e^{-\lambda_1 x});$$

$$\frac{dV_2^*(x)}{dx} = \frac{c^+}{d}x - \frac{q_2}{q_1}\frac{J^*}{d}(1 - e^{-\lambda_1 x}) - \frac{J^*}{d}\left(1 + \frac{q_2}{q_1}\right).$$

Lemma 3 *The value functions V_1^* and V_2^* are continuously differentiable.*

Proof. The only point at which the continuity of the derivatives of V_1^*, and V_2^* is not established is at $x = 0$. This can be verified using the identity $f_3(z_2^*) = 0$. □

Proposition 7 *The set of conditions*

$$\frac{c^-}{q_2} < c_v \leq \frac{c^+}{q_1} \tag{17}$$

is sufficient for the convexity of V_1^*, and V_2^*.

Proof. The convexity of V_1^*, and V_2^* on $(-\infty, 0)$ is already established. On the interval $[0, \infty)$ we have:

$$\frac{d^2V_1(x)}{dx^2} = \lambda_1 \frac{J^*}{d}(1 - e^{-\lambda_1 x}) + \frac{1}{d}[c^+ - c_v q_1$$
$$+ \frac{r_1 - d}{d} q_2(c_v - \frac{c^-}{q_2}(1 - \frac{q_1 + q_2}{r_1 - d} z_2))];$$

$$\frac{d^2V_2(x)}{dx^2} = \frac{q_2}{q_1}\frac{J^*}{d}\lambda_1 e^{-\lambda_1 x} + \frac{c^+}{d}.$$

It is easy to see that $\frac{d^2V_2(x)}{dx^2}$ is positive. As for $\frac{d^2V_1(x)}{dx^2}$, note that the first term is clearly positive. We also know that $c_v < \frac{c^+}{q_1}$ and $\frac{c^-}{q_2} < c_v$ imply $c_v \geq \frac{c^-}{q_2}(1 - \frac{q_1+q_2}{r_1-d}z_2^*)$ (Lemma 2). Therefore the second term is also positive. Hence $\frac{d^2V_1(x)}{dx^2} > 0$ on $[0, \infty)$ and the proof is complete. □

Theorem 5 *Under conditions (17) there exists a hedging point policy with zero inventory level ($z_1^* = 0$) and negative backlog level ($z_2^* < 0$) that is optimal among all feasible policies.*

Proof. From our previous discussions it follows that under conditions (17) there exist a hedging point policy for which $z_1^* = 0$, $z_2^* < 0$, and the associated value functions are convex and continuously differentiable. To show that this policy is optimal among all feasible policies it is sufficient to show that it satisfies the HJB equations. The proof is almost identical to the proof of Theorem 3 and is based on the fact that $\frac{dV_1^*}{dx}$ changes sign from negative to positive at $x = z_1^* = 0$, $\frac{dV_1^*}{dx} - c_v$ changes sign from negative to positive at $x = z_3^*$, and $\frac{dV_2^*}{dx} - c_v$ changes sign from negative to positive at $x = z_2^*$. □

We point out that the sufficient condition $c_v \leq c^+/q_1$ used here for the optimality of the hedging point policy with $z_1^* = 0$) and $z_2^* < 0$ can be replaced two weaker conditions

$$f_1(0) \geq 0 \qquad (18)$$

$$dp_2[\rho(\frac{c^-}{q_2} - c_v) - (\frac{c^+}{q_1} - c_v)] \leq 0. \qquad (19)$$

(That is to say $c_v \leq c^+/q_1$ implies (18) and (19).) In fact, it can be shown that (19) is also a necessary condition for the convexity.

4.3 Zero backlog, nonzero inventory ($z_1^* > 0$, $z_2^* = 0$)

The arguments used for establishing results in this case are very similar to those used in the previous case. A necessary condition for the existence of an optimal hedging point policy with non-negative inventory hedging level and zero backlog is that this hedging point policy be optimal among all hedging point policies of the same type. A necessary condition for the latter to be true is

$$\frac{\partial J(z_1, 0)}{\partial z_1}\bigg|_{z_1=z_1^*} = 0.$$

Taking derivative of $J(z_1, 0)$ with respect to z_1, simplifying, and setting it equal to zero yields the following equation:

$$f_4(z_1) = \frac{1}{\lambda_0}[c^+\lambda_0 z_1 + c^+\rho(e^{-\lambda_0 z_1} - 1) + \frac{d(r_1 - d)}{q_1 + q_2}\lambda_0^2(\frac{c^+}{q_1} - c_v)] = 0.$$

Proposition 8 *A necessary and sufficient condition for the existence of a positive root for $f_4(z_1) = 0$ is*

$$c_v > \frac{c^+}{q_1};$$

such a negative root is unique.

Proof. Let $g_2(z_1) = \lambda_0 f_4(z_1)$. The roots of f_4 and g_2 coincide so we can focus on the function g_2. Note that $\frac{d^2 g_2}{dz_1^2}(z_1) = c^+\rho\lambda_0^2 e^{-\lambda_0 z_2} > 0$, hence g_2 is convex. Moreover, it is easy to verify that the minimizer of g_2 is at $z_1^m = \frac{\ln\rho}{\lambda_0}$; note that $\rho < 1 \Leftrightarrow \lambda_0 < 0$, therefore $z_1^m > 0$. Hence a necessary and sufficient condition for the existence of a negative root for g_2 is $g_2(0) = \frac{d(r_1-d)}{q_1+q_2}\lambda_0^2(\frac{c^+}{q_1} - c_v) < 0$, which is equivalent to $c_v > \frac{c^+}{q_1}$. Clearly, in this case, the negative root of g_2, and hence of f_4, is unique. □

The following corollary is one of the main results of this section:

Theorem 6 *A necessary condition for the existence of a hedging point policy π^{z*} with $z_1^* > 0, z_2^* = 0$, that is optimal among all feasible policies, is*

$$c_v > \frac{c^+}{q_1}. \tag{20}$$

A useful result about the hedging level z_1^* which will be needed in what follows is:

Lemma 4 *Assume $c_v > \frac{c^+}{q_1}$, and let z_1^* be as defined above. Then*

$$\frac{q_1 + q_2}{q_1 d} c^+ z_1^* < c_v - \frac{c^+}{q_1}. \tag{21}$$

Proof. Assume $\lambda_0 > 0$. Using the inequality $(e^{-\lambda_0 x} - 1) > -\lambda_0 x$ in the equation $f_4(z_1^*) = 0$, we get

$$c^+(1-\rho)z_1^* < \frac{d(r_1 - d)}{q_1 + q_2} \lambda_0 (c_v - \frac{c^+}{q_2}).$$

The result follows immediately from the above inequality. The case of $\lambda_0 < 0$ can be proved similarly. □

Let z_1^* and 0 denote the optimal hedging levels (when they exist) and let J^* denote the corresponding cost. J^* is given by

$$J^* = c^+ z_1^* + \frac{dc^+}{q_1 + q_2} \tag{22}$$

The above cost function J^* and that for the case $z_1^* > 0, z_2^* < 0$ (equation (12), J^* expressed in terms of z_1^*) are identical in form. Hence, using an argument similar to what we gave in the previous section, it can be shown that the value functions V_1^* and V_2^* are also similar in form on the interval $[0, \infty)$. Therefore, V_1^* and V_2^* are convex and continuously differentiable on the interval $[0, \infty)$. We focus on the interval $(-\infty, 0)$.

For $z_3 \leq x \leq 0$

$$\frac{dV_1^*(x)}{dx} = \frac{c^-(q_1 + q_2)}{(r_1 - d)(r_2 - d)} \frac{q_1}{q_2} x + \frac{q_1}{r_1 - d}[\frac{J^* - c_v d}{q_2} + \frac{\delta_2 J^*}{\lambda_2}$$

$$\begin{aligned}
&\quad + \frac{\delta_2 c^-}{\lambda_2^2}]e^{\lambda_2 x} - \frac{q_1}{r_1-d}[\frac{\delta_2 J^*}{\lambda_2} + \frac{\delta_2 c^-}{\lambda_2^2}] \\
&\quad + \frac{q_1}{r_1-d}\frac{c_v r_2}{\lambda_2(r_2-d)}[e^{\lambda_2 x}-1] + \frac{J^*}{r_1-d}; \\
\frac{dV_2^*(x)}{dx} &= \frac{c^-(q_1+q_2)}{(r_1-d)(r_2-d)}\frac{q_1}{q_2}x - \frac{q_2}{r_2-d}[\frac{J^*-c_v d}{q_2} + \frac{\delta_2 J^*}{\lambda_2} \\
&\quad + \frac{\delta_2 c^-}{\lambda_2^2}]e^{\lambda_2 x} + \frac{q_2}{r_2-d}[\frac{\delta_2 J^*}{\lambda_2} + \frac{\delta_2 c^-}{\lambda_2^2}] \\
&\quad - \frac{q_2}{r_2-d}\frac{c_v r_2}{\lambda_2(r_2-d)}[e^{\lambda_2 x}-1] + \frac{J^*}{r_2-d} - \frac{c_v r_2}{r_2-d},
\end{aligned}$$

where $\delta_2 = 1/(r_1-d)+1/(r_2-d)$. As will be shown shortly, under conditions

$$\frac{c^+}{q_1} < c_v \leq \frac{c^-}{q_2} \qquad (23)$$

we can uniquely specify an appropriate z_3^* and hence completely specify the policy π^{z^*}. Then we have

Proposition 9 *Under conditions (23) the value functions V_1^* and V_2^* are convex and continuously differentiable.*

Proof. We first complete the definition of the policy π^{z^*}. On the interval $[z_3, 0)$, we have

$$\frac{d^2 V_1^*(x)}{dx^2} = \frac{c^-(q_1+q_2)}{(r_1-d)q_2+(r_2-d)q_1}(1-e^{\lambda_2 x})$$
$$+ \frac{q_1}{r_1-d}\frac{c^+}{dq_1(1-\rho)}(r_1(1-e^{\lambda_2 z_1^*}) + d(1-\rho)e^{\lambda_2 z_1^*})e^{\lambda_2 x}$$
$$+ \frac{c^-}{r_1-d}e^{\lambda_2 x}$$

$$\frac{d^2 V_2^*(x)}{dx^2} = \frac{q_2}{(r_2-d)\rho}(c_v - (\frac{c^+}{q_1} + \frac{q_1+q_2}{dq_1}c^+ z_1^*))e^{\lambda_2 x} + \frac{q_2}{r_2-d}(\frac{c^-}{q_2}-c_v)e^{\lambda_2 x}$$

It is easy to see that $\frac{d^2 V_1^*(x)}{dx^2} > 0$ and under conditions (23) $\frac{d^2 V_2^*(x)}{dx^2} > 0$. Hence, the value functions are convex on $[z_3, 0)$ and therefore $\frac{dV_1^*(x)}{dx}$ is strictly increasing. Note that $\lim_{x \to -\infty} \frac{dV_1^*(x)}{dx} = -\infty$. On the other hand it can be shown that $\frac{dV_1^*(0)}{dx} > -c_v$, hence there exists a unique z_3^* such that $\frac{dV_1^*(0)}{dx} = -c_v$. We have now completely defined the policy π^{z^*}. To verify that the value functions are continuously differentiable for this policy we only need to consider the point $x = 0$; as in the other cases, the continuity of the derivatives follows from the characterizing equation for the optimal hedging levels ($f_4(z^*) = 0$ in this case). Convexity of the value functions has already been established on $[z_3^*, \infty)$; convexity on $(-\infty, z_3^*)$ follows from arguments similar to those given in section 4.1. □

Theorem 7 *Under conditions (23) there exists a hedging point policy with positive inventory level ($z_1^* > 0$) and zero backlog level ($z_2^* = 0$) that is optimal among all feasible policies.*

Proof. The proof is identical to the corresponding theorem in the previous section. □

Similar to the second case, the condition $c_v \leq c^-/q_2$ used here for the optimality of the hedging point policy with $z_1^* > 0$ and $z_2^* = 0$ can be replaced by two weaker conditions

$$f_2(0) \geq 0, \qquad (24)$$

$$dp_2[\rho(\frac{c^-}{q_2} - c_v) - (\frac{c+}{q_1} - c_v)] \geq 0. \qquad (25)$$

(i.e., $c_v \leq c^-/q_2$ implies (24) and (25).) Furthermore, it can be shown that (25) is also a necessary condition for the convexity.

4.4 Zero inventory, zero backlog ($z_1^* = 0$, $z_2^* = 0$)

With no inventory and no backlog, the only cost in this case is the cost of purchasing extra capacity. This cost is given by

$$J^* = c_v d \frac{q_1}{q_1 + q_2}.$$

We can prove the first basic result of this subsection immediately.

Theorem 8 *The set of conditions*

$$c_v \leq \frac{c^+}{q_1}, \quad c_v \leq \frac{c^-}{q_2} \qquad (26)$$

is necessary for the hedging point policy with $z_1^* = 0$, $z_2^* = 0$ to be optimal among all feasible policies.

Proof. For this policy to be optimal it is necessary that $J(0,0) \leq J(z_1,0)$ for all $z_1 > 0$ and $J(0,0) \leq J(0,z_2)$ for all $z_2 < 0$. Therefore, in light of Propositions 6, it is necessary that f_3 have no negative roots, and, in light of Propositions 7, it is necessary that f_4 have no positive roots; these latter conditions are equivalent to (as proved in Propositions 6 and 7) $c_v \leq \frac{c^+}{q_1}$ and $c_v \leq \frac{c^-}{q_2}$. $\square$

The value functions associated with the above policy can be calculated as follows.

For $0 \leq x$

$$\frac{dV_1^*(x)}{dx} = \frac{q_1 + q_2}{d^2}[c^+ \frac{1}{\lambda_1}x - c_v q_1 \frac{1}{\lambda_1^2}(1 - e^{-\lambda_1 x})];$$

$$\frac{dV_2^*(x)}{dx} = \frac{q_1 + q_2}{d^2}[c^+ \frac{1}{\lambda_1}x + c_v q_2 \frac{1}{\lambda_1^2}(1 - e^{-\lambda_1 x})].$$

Taking derivative we have

$$\frac{d^2V_1^*(x)}{dx^2} = \frac{q_1}{d}[(\frac{c^+}{q_1} - c_v) + c_v(1 - e^{-\lambda_1 x})];$$

$$\frac{d^2V_2^*(x)}{dx^2} = \frac{q_2}{d}[c_v e^{-\lambda_1 x} + \frac{c^+}{q_2}].$$

From the above equations it can be simply verified that $\frac{d^2V_2^*(x)}{dx^2}$ is positive and that a sufficient condition for $\frac{d^2V_1^*(x)}{dx^2}$ to be positive is

$$c_v < \frac{c^+}{q_1}. \tag{27}$$

For $z_3 \leq x < 0$

$$\frac{dV_1^*(x)}{dx} = \frac{c^-(q_1+q_2)}{(r_1-d)(r_2-d)}[\frac{1}{\lambda_2}x + \frac{1}{\lambda_2^2}(1-e^{\lambda_2 x})] - \frac{c_v q_1 + c^-}{(r_1-d)}\frac{1}{\lambda_2}(1-e^{\lambda_2 x})$$

$$\frac{dV_2^*(x)}{dx} = \frac{c^-(q_1+q_2)}{(r_1-d)(r_2-d)}\frac{1}{\lambda_2}x - \frac{c^-(q_1+q_2)}{(r_2-d)^2}\frac{q_2}{q_1}\frac{1}{\lambda_2^2}(1-e^{-\lambda_2 x})$$

$$+ \frac{q_2}{q_1}\frac{c_v q_1 + c^-}{(r_2-d)}\frac{1}{\lambda_2}(1-e^{-\lambda_2 x}) - c_v$$

Taking derivative we have

$$\frac{d^2V_1^*(x)}{dx^2} = \frac{c^-(q_1+q_2)}{(r_1-d)(r_2-d)\lambda_2}(1-e^{\lambda_2 x}) + \frac{c_v q_1 + c^-}{(r_1-d)}e^{\lambda_2 x};$$

$$\frac{d^2V_2^*(x)}{dx^2} = \frac{c^-(q_1+q_2)}{(r_1-d)(r_2-d)\lambda_2}(1-e^{\lambda_2 x}) + \frac{q_2}{(r_2-d)}(\frac{c^-}{q_2} - c_v)e^{\lambda_2 x}.$$

From the above identities it can be simply verified that $\frac{d^2V_1^*(x)}{dx^2}$ is positive and that a sufficient condition for $\frac{d^2V_2^*(x)}{dx^2}$ to be positive is

$$c_v < \frac{c^-}{q_2}. \tag{28}$$

If the above two conditions (27) and (28) are satisfied, as in the previous sections, we can define a z_3^* such that $\frac{dV_1^*(z_3^*)}{dx} = -c_v$. For the resulting policy we have:

Proposition 10 *The set of conditions (26) is necessary and sufficient for the value functions V_1^*, and V_2^* to be convex and continuously differentiable.*

Proof. Continuity of the derivatives follows directly from the above identities. We already showed that under the conditions of the proposition, the value functions are convex on $[z_3^*, \infty)$. The proof of convexity for values below z_3^* is similar to that given in section 4.1 for the same region.

We now can give the main result of this subsection:

Theorem 9 *Under conditions (26) there exists a hedging point policy with zero inventory level ($z_1^* = 0$) and zero backlog level ($z_2^* = 0$) that is optimal among all feasible policies.*

Proof. The proof is identical to the corresponding theorem in section 4.2. □

5 Conclusion

We studied a flow control problem for a system in which it is possible to obtain extra capacity at additional cost. We set up the problem in a general setting. For systems with one part-type and two machine states, we studied the hedging point policies with three hedging levels. We derived necessary and sufficient conditions under which such a hedging point policy is optimal for four different cases, depending on whether the hedging levels are zero or not. Also, the optimal values of the three hedging levels were obtained for these cases. However, the conditions we established do not cover all parameter ranges, and can be improved. This is a future research direction.

References

[1] R. Akella and P.R. Kumar, "Optimal Control of Production Rate in a Failure Prone Manufacturing System," *IEEE Transactions on Automatic Control,* Vol. 31, No. 2, pp. 116-126, February 1986.

[2] T. Bielecki and P.R. Kumar, "Optimality of Zero-Inventory Policies for Unreliable Manufacturing Systems," *Operations Research,* Vol. 36, No. 4, pp. 532-541, July-August 1988.

[3] M. Caramanis and A. Sharifnia, "Design of Near Optimal Flow Controllers for Flexible Manufacturing Systems," *Proceedings of the 3th ORSA/TIMS special Interest Conference on FMSA,* Cambridge, MA, K.E. Steche and R. Suri eds. Elsevier Science Publishers B.V., Amsterdam.

[4] M. Caramanis and G. Liberopoulos, "Perturbation Analysis for the Design of Flexible Manufacturing Systems Flow Controllers," *Operations Research* , Vol. 40, No. 6, pp. 1107-1125, Nov.-Dec. 1992.

[5] S. Gershwin, *Manufacturing Systems Engineering,* Prentice Hall, 1993.

[6] S. Gershwin, Personal Communication, 1994.

[7] S.B. Gershwin, R. Akella, and Y.F. Choong, "Short-Term Production Scheduling of an Automated Manufacturing Facility," *IBM Journal of Research and Development,* Vol. 29, pp. 392-400, July 1985.

[8] J.Q. Hu, "Production Rate Control for Failure Prone Production With No Backlog Permitted," *IEEE Transactions on Automatic Control,* Vol.40, No.2, pp. 291-295, 1995.

[9] J.Q. Hu, "A Queueing Approach to Manufacturing Flow Control Models," *Proceedings of the 34th IEEE Conference on Decision and Control,* December 13-15, 1995.

[10] J.Q. Hu, P. Vakili, and G.X. Yu, "Optimality of Hedging Point Policies in the Production Control of Failure Prone Manufacturing Systems," *IEEE Trans. on Automatic Control,* Vol.39, No.9, pp.1875-1880, 1994.

[11] J.Q. Hu and D. Xiang, "The Queueing Equivalence to a Manufacturing System with Failures," *IEEE Transactions on Automatic Control,* **38,** pp. 499-502, March, 1993.

[12] J.Q. Hu and D. Xiang, "Structure Properties of Optimal Production Controllers of Failure Prone Manufacturing Systems," *IEEE Transactions on Automatic Control,* Vol.39, No.3, pp. 640-643, 1994.

[13] J.Q. Hu and D. Xiang, "Optimal Control for Systems with Deterministic Cycles," *IEEE Transactions on Automatic Control,* Vol.40, No.4, pp. 782-786, 1994.

[14] J.Q. Hu and D. Xiang, "Monotonicity of Optimal Flow Control for Failure Prone Production Systems," *Journal of Optimization Theory and Applications,* Vol. 86, No.1, pp. 57-71, July 1995.

[15] J. Lehoczky, S.P. Sethi, H.M. Soner, and M.I. Taksar, "An Asymptotic Analysis of Hierarchical Control of Manufacturing Systems under Uncertainty," *Mathematics of Operations Research,* **16,** pp. 594, 1991.

[16] J.G. Kimemia and S.B. Gershwin, "An Algorithm for the Computer Control of Production in Flexible Manufacturing Systems," *IIE Transactions,* Vol. 15, pp. 353-362, December 1983.

[17] G. Liberopoulos and M. Caramanis, "Production Control of Manufacturing Systems with Production Rate Dependent Failure Rates," *IEEE Trans. on Automatic Control,* Vol. 39, No. 4, pp.889-895, 1994.

[18] G. Liberopoupos and J.Q. Hu, "On the Order of Optimal Hedging Points in a Class of Manufacturing Flow Control Models," *IEEE Transactions on Automatic Control,* Vol.40, No.2, pp. 282-286, 1995.

[19] G.J. Olsder and R. Suri, "The Optimal Control of Parts Routing in a Manufacturing System with Failure Prone Machines," *Proceedings of the 19th IEEE Conference on Decision and Control*, 1980.

[20] R. Rishel, "Dynamic Programming and Minimum Principles for Systems with Jump Markov Distributions," *SIAM Journal of Control*, Vol. 29, No. 2, February 1975.

[21] S. Sethi, H. M. Soner, Q. Zhang, and J. Jiang, "Turnpike Sets and their Analysis in Stochastic Production Planning Problems," *Math. OR*, Vol. 17, pp. 932–950, 1992.

[22] S.P. Sethi and Q. Zhang, *Hierarchical Decision Making in Stochastic Manufacturing Systems*, Birkhauser, Boston, 1994.

[23] A. Sharifnia, "Optimal Production Control of a Manufacturing System with Machine Failures," *IEEE Transactions on Automatic Control*, Vol. AC-33, No. 7, pp. 620-625, July 1988.

[24] G.X. Yu and P. Vakili, "Control of Manufacturing Systems with Production Dependent Machine Failures," in *Proceedings of 31st Annual Allerton Conference on Communication, Control, and Computing*, Allerton, IL, 1993.

LEI HUANG, JIAN-QIANG HU, AND PIROOZ VAKILI
MANUFACTURING ENGINEERING DEPARTMENT
BOSTON UNIVERSITY
44 CUMMINGTON STREET
BOSTON, MA 02215

Optimal Control of Manufacturing Systems with Buffer Holding Costs

James R. Perkins and R. Srikant

Abstract

In this paper, we examine the optimal control of a fluid flow model for manufacturing systems. The objective is to minimize inventory and shortfall costs. Under the assumption that the machines are reliable, we discuss and extend the method of solution used by Perkins and Kumar [31]. This approach exploits the structural properties of controls which perform well, and shows that an optimal control is determined by solving a series of quadratic programming problems having linear inequality constraints. Equivalently, the optimal control can be expressed by the description of a series of linear switching surfaces, which can be computed off-line (for a given manufacturing system).

In the second part of the paper, we assume the machines are failure-prone. Motivated by the results when the machines are reliable, we consider the application of linear switching curve controls. Under such controls, for systems consisting of two buffers and two machine states, we show that the steady-state probability densities of the buffer levels satisfy a set of coupled hyperbolic partial differential equations. Perkins and Srikant [33] use transform techniques to explicitly determine these probability densities. For a certain class of policies, called prioritized hedging point policies, these results straightforwardly extend to an arbitrary number of part-types. We then discuss the application of these results to the determination of optimal and near-optimal controls.

1 Introduction

This paper discusses and extends recent results achieved towards the optimal control of manufacturing systems. We consider two subproblems, one deterministic and one stochastic (with machine failures). For the deterministic problem, an optimal control is obtained. The form of this control is then used to obtain a set of controls for the stochastic problem. Then, the steady-state behavior of these near-optimal controls is explicitly determined.

The first subproblem to be considered consists of a re-entrant manufacturing system with buffer holding costs. It is assumed that the machines are reliable and the system parameters are deterministic and constant. The performance objective is the minimization of work-in-process (WIP) and backlog costs. However, it should be noted that other first order criteria, such as minimization of delay or maximization of throughput, can be incorporated by appropriate choice of the system cost parameters.

In an optimal control framework, the formulation of this problem yields a minimization problem having both control and state constraints. The solution of this type of problem is known to be extremely difficult. In general, even the existence of an optimal control has not been shown.

We begin by discussing and extending the results of Perkins and Kumar [31] for a flow shop, i.e., a system in which there is no re-entrant flow. The method of solution constructively exploits the structural properties of controls which perform well, and thereby limits the control space to be searched. This approach significantly reduces the complexity of the problem. In fact, the optimal scheduling problem is reduced to the determination of optimal switching curves, given by linear hyperplanes, which can be easily generated off-line by solving a series of quadratic programming problems with linear inequality constraints.

There are several papers which are closely related to this first subproblem. Hajek and Ogier [18] consider a problem in which the objective is to empty a fluid model for a communications network in minimum time. However, in their paper, there is no drainage due to demand, and hence no negative buffer levels. Avram, Bertsimas, and Ricard [3] propose a near-optimal control for a similar problem. Also, Chen and Yao [11] consider a push model with buffer costs. They show that when a unique globally optimal allocation (a control which minimizes a cost criterion over all time horizons) exists, it can be determined through a myopic procedure. In addition, they show that there are cases when such an allocation does not exist. It can be shown that a necessary condition for a globally optimal control to exist is that the control must empty the buffers of the system in minimum time. In contrast, the solution to the proposed subproblem (which is believed to always exist) will meet this more stringent objective only in special instances.

More recently, Sethi and Zhou [35] determine the optimal control for a two-machine flowshop. They then use this control to form an asymptotically optimal control for the case in which the machines are unreliable. Connolly [12] considers the problem of when to schedule set-ups in a manufacturing system. As part of the analysis, the optimal control is determined for a two part-type system. More general cost structures are examined in Wu and Perkins [42].

The second subproblem considers the case in which a machine serving multiple buffers is unreliable. In a general re-entrant manufacturing system, if a machine is a bottleneck for a particular buffer, then it may be necessary to keep some safety stock in the nearby buffers, both to allow for processing on that buffer when the machine is up, and to allow the downstream buffers to provide material when the machine is down (see, for example, Akella and Kumar [1] and Bielecki and Kumar [6]).

Thus, the results of the second subproblem are incorporated to determine a local control. It should be noted that it is assumed that the system contains a relatively small number of buffers of this type. For many manufacturing systems, this is a reasonable assumption. For example, in the three systems considered by Wein [39] and Lu and Kumar [25, 26], there are one, two, and four bottleneck machines, respectively, out of a total of 24 service stations and 172 buffers. If there are a large number of bottleneck machines with unreliability of the same order of magnitude as the system dynamics, then the system is inherently volatile. In this case, the input control and detailed scheduling of machines become a second-order concern.

As shown in Perkins and Kumar [30] for a two machine example, if the machine states occur as a continuous-time Markov chain and the imposed control induces stability in the system, then the resultant steady-state probability densities for the buffer levels will satisfy a system of linear partial differential equations with boundary constraints in the form of ordinary differential equations.

Perkins and Srikant [33] consider a single machine system producing multiple part-types. This leads to a similar set of coupled partial differential equations. In fact, for any control which regulates production based on the comparison of a linear combination of the buffer levels to a threshold value, the steady-state probability densities will satisfy such a set of linear, coupled differential equations. A solution to these coupled hyperbolic partial differential equations is of interest both to the mathematical community and in the field of manufacturing, where little progress has been made beyond the single machine system.

This subproblem is a special case of the general problem posed by Kimemia and Gershwin [21] in their pioneering paper. Heuristic arguments in that paper based on the Hamilton-Jacobi-Bellman (HJB) sufficient conditions and the works of Rishel [34] and Tsitsiklis [38] suggest that the optimal solution is a stationary, feedback solution with a renewal point (i.e., a hedging point) for the buffer level stochastic process. When the buffer level is at the hedging point, each part-type is produced at a rate equal to its demand rate. When the buffer levels are not at the hedging point, then the production rates operate at the boundary of the control constraint set. The buffer space is divided into regions such that, inside each region, the production rates for all of the part-types are constant, and the production rates change at the region boundaries, also known as "switching curves."

The analytical determination of the optimal hedging point and the optimal switching curves have been elusive except in the case of the single-machine, single part-type, linear cost function problem which was completely solved by Bielecki and Kumar [6]. (See also Glasserman [17] and Sharifnia [36], for the case of multiple states on a single machine.) A partial characterization of the structure of the optimal control policy for a multiple part-type problem has been obtained recently by Srivatsan [37].

Several sub-optimal solutions have been suggested for the general problem in the literature. Kimemia and Gershwin [21] and Gershwin, Akella, and Choong [16] proposed a quadratic approximation to the differential cost-to-go function of the HJB equation. Caramanis and Liberopoulos [8] used the quadratic approximation to estimate the hedging point by combining perturbation analysis with simulation. Experiments by Boukas and Haurie [7] suggest that the quadratic approximation is close to optimal. Another method for estimating the hedging point, due to Caramanis and Sharifnia [9], is the decomposition of a multiple part-type problems into single part-type problems by inscribing hypercubes in the capacity constraint set.

The rest of the paper is organized as follows. In Section 2, we introduce the fluid flow model. In Section 3, we assume that machines are reliable. We describe and extend the method used by Perkins and Kumar [31] to obtain an optimal control for the system. Section 4 considers the case in which there is machine unreliability. Under the linear switching curve (LSC) scheduling policy, the steady-state probability density of the buffer levels is obtained. We then generalize the results to machines that produce more than two part-types. Concluding remarks are presented in Section 5.

2 The Fluid Flow Model

Consider a re-entrant manufacturing system, of the type shown in Figure 1, which produces a single part–type on R machines $M_0, \ldots, M_{R-1}$. Assume that parts follow a fixed route through the system, requiring processing at machines $M_{\lambda_0}, M_{\lambda_1}, \ldots, M_{\lambda_{L-1}}$, where $\lambda_i \in \{0, 1, \ldots, R-1\}$.

This model assumes that the part flow is fluid, with the implicit assumption that we are modeling a high volume manufacturing system. As is well-documented in the literature (see, for example, Chen and Mandelbaum [10], Harrison and Wein [19], and Wein [40, 41]), a fluid flow model often can effectively describe the heavy traffic behavior of discrete part queueing networks.

It is important to point out that the assumption of fluid part flow is controversial. However, it is not the purpose of this paper to explore this issue. We will suffice it to say that, for real-world systems which do not precisely adhere to our modeling assumptions, the results derived herein provide important qualitative guidelines for efficient controls.

This model also assumes that machine set-up times either can be included in the processing times or are negligible compared to the system dynamics. For treatment of systems in which set-up times are nonnegligible, the reader is referred to Perkins [27], Perkins and Kumar [29] and Kumar [22], and the references contained therein.

Assume that parts at the machine queue in buffers, while awaiting processing. Label these buffers $\{b_i : 0 \leq i \leq L\}$ in increasing order, where buffer b_i is served by machine M_{λ_i}. It is assumed that b_0 is an infinite reservoir of raw parts and that production responds to demand. In order to model this, the output buffer b_L is depleted at a constant rate d, due to exogenous demand for the finished product, and replenished by parts exiting machine $M_{\lambda_{L-1}}$.

Let $x_i(t)$ be the level of buffer b_i at time t. Thus, $x_L(t)$ represents the *net inventory level* of the finished commodity at time t. Note that $x_L(t)$ may be positive or negative, depending on whether there is a surplus or a shortfall. However, the remaining physical buffers all must contain a nonnegative amount of material. Thus, $x_i(0) \geq 0$ for $1 \leq i \leq L-1$.

Define $u_j(t)$ to be the processing rate of material in buffer b_j at time t. Hence, $\{u_0(t), t \geq 0\}$ controls the release of material into the system. The dynamics of $x_i(t)$, the level of buffer b_i, satisfy the equation

$$x_i(t) := x_i(0) + \int_0^t [u_{i-1}(\sigma) - u_i(\sigma)] d\sigma.$$

Let μ_j be the maximum production rate for machine M_j. Then, a necessary capacity condition for stability is $\mu_i > d$ for all machines M_i.

A control $u(t)$ is *feasible* if it meets the following conditions:

(i) $u(t) = (u_0(t), \ldots, u_{L-1}(t))$ is a measurable function.

(ii) $0 \leq \sum_{\{i:\lambda_i=j\}} u_i(t) \leq \mu_j$ for all $t \geq 0$.

(iii) For $1 \leq i \leq L-1$, if $x_i(t) = 0$, then $u_i(t) \leq u_{i-1}(t)$.

Given the model described, suppose that a unit of material in buffer b_i incurs a holding cost of c_i units per unit time, for $1 \leq i \leq L-1$. Note that no costs accrue for material in the input buffer b_0. For the output virtual buffer b_L, each unit of inventory incurs a cost of c_L^+ units per unit time, while each unit of shortfall incurs a cost of c_L^- units per unit time. Assume $c_0 = 0, c_i \geq 0$ for $0 \leq i \leq L-1$ and $c_L^+, c_L^- \geq 0$.

We first consider the case in which the machines are reliable.

3 Optimal Control of Reliable Pull Manufacturing Systems

The goal of the deterministic scheduling problem is to determine the vector production rate $u(t)$, from the set of feasible controls, so as to minimize the infinite horizon total cost

$$\int_0^{+\infty} \sum_{i=1}^{L} c_i x_i(t) dt. \tag{1}$$

Let $J(u)$ denote the cost of a control u.

As long as the system has adequate capacity to meet demand, it will eventually empty. Thus, the objective is to determine the optimal control for the production rate at each of the machines in the system, in order to minimize the total buffer holding cost incurred while emptying all material from the internal buffers of the system and recouping any shortfall.

Consider the model for a re-entrant manufacturing line, as described in Section 2. The primary focus will be on the case in which the holding costs are nondecreasing, i.e., $0 =: c_0 \leq c_1 \leq \ldots \leq c_{L-1} \leq c_L^+$. This assumption is natural in semiconductor manufacturing systems, as well as many other types of systems, since holding costs typically increase as the "value added" increases. The form of the optimal control is particularly easy to deduce when $x_L(0) \geq 0$, i.e., when there is a surplus of finished product.

3.1 Initial Surplus Case: $x_L(0) \geq 0$

Note that when the level of the output virtual buffer is initially nonnegative,

$$T = \frac{\sum_{i=1}^{L} x_i(0)}{d} \tag{2}$$

is the minimum possible value of the system clearing time. Thus, if we consider an infinitesimal amount of material, it will leave the system at the same time, regardless of the control implemented. Since holding costs increase downstream, it follows that an optimal policy for each machine is for it to remain idle until all of the buffers downstream from it are empty, and then to process material at rate d.

Theorem 3.1 *Consider a re-entrant line, such as the system shown in Figure 1. Assume that the part–type visits the buffers in the order $b_0, \ldots, b_L$. Then, if $x_L(0) \geq 0$, an optimal control at buffer b_i is*

$$u_i(t) = \begin{cases} d & \text{if } x_j(t) = 0 \text{ for all } j > i, \\ 0 & \text{otherwise.} \end{cases} \tag{3}$$

If all of the buffers, including the output virtual buffer (but excluding b_0) are empty, then the optimal solution is to *pipeline* material at rate d, incurring no additional cost. Thus, by the Principle of Optimality (see Bellman [4]), after a system clearing time, T, the optimal solution is $u_i(t) \equiv d$, for $0 \leq i \leq L-1$ and all $t \geq T$.

3.2 Initial Shortfall Case: $x_L(0) < 0$

This case is considerably more complex. There is a tradeoff between producing and not producing. By producing, a machine is able to lower shortfall costs. However, production causes a machine to transfer its buffer contents to the next downstream machine, which incurs higher holding costs. Thus, while it is important to keep material in the earlier buffers as long as possible because of the increasing costs, it is also important to "put out the fire," caused by the last buffer level being negative. The optimal policy must strike a balance between these two conflicting objectives.

We first examine the approach used by Perkins and Kumar [31] to solve this problem for a single part-type flow shop, i.e., no re-entrant flow is allowed. A constructive approach is used, which iteratively restricts the class of controls which perform well, to show that an optimal control does indeed exist, and that it can be characterized by N "deferral" times, where N is no more than the number of machines. The optimal control problem is then determined by solving a set of N quadratic programming problem with linear inequality constraints.

Consider the manufacturing system shown in Figure 2. It consists of L machines $\{M_0, M_1, \ldots, M_{L-1}\}$ in tandem. The system produces a single part–type which visits the machines in the order $M_0, M_1, \ldots, M_{L-1}$. Parts leaving machine M_i flow into a buffer b_{i+1}, where they are held for processing by machine M_{i+1}. Suppose that machine M_i can process parts at a rate up to μ_i parts per unit time.

The set of machines $\mathcal{M} := \{M_0, M_1, \ldots, M_{L-1}\}$ is partitioned into $N (\leq L)$ *sections* as follows:

$$S_1 = \{M_0, \ldots, M_{s_1}\}, S_2 = \{M_{s_1+1}, \ldots, M_{s_2}\}, \ldots, S_N = \{M_{s_{N-1}+1}, \ldots, M_{s_N}\} \tag{4}$$

where $s_N := L - 1$, and s_{i-1} is recursively defined by

$$s_{i-1} := \max\{0 \leq j < s_i : \mu_j < \mu_{s_i}\}. \tag{5}$$

Thus, in each section S_i, the bottleneck is the most downstream machine M_{s_i} with the smallest processing rate μ_{s_i}. For convenience, we define

$$\bar{\mu}_j := \mu_{s_i} \text{ for all of the machines } M_j \text{ in section } S_i. \tag{6}$$

In order to facilitate the analysis, it is necessary to introduce some notation. Define T^u to be the first time at which the output virtual buffer b_L no longer has a shortfall, when the control $u(t)$ is implemented. That is,

$$T^u := \inf\{\sigma \geq 0 : x_L(\sigma) \geq 0\}. \tag{7}$$

By the Principle of Optimality, the optimal solution for all $t \geq T^u$ is given by Theorem 3.1. Thus, we are interested in determining an optimal control for $t < T^u$.

Clearly, as long as there is a shortfall, the final machine will process material as quickly as possible. Thus, given an initial shortfall $x_L(0) < 0$ and a feasible control u, we can determine the number of buffers that are needed to eliminate the shortfall of the output virtual buffer. Define

$$i(u) := \max\{0 \leq i \leq L-1 : \sum_{j=i}^{L-1} x_j(0) \geq -x_L(0) + T^u d \text{ under the control } u\}. \tag{8}$$

We will call $i(u)$ the *shortfall erasing buffer* under control u. Thus, the control $u(t)$ uses the material which is initially in buffers $b_{i(u)}, \ldots, b_{L-1}$ to erase the shortfall.

Perkins and Kumar [31] iteratively restrict the class of controls, until reaching a class they call $\mathcal{F}_8$. A control $u \in \mathcal{F}_8$ possesses the following properties:

- The control must be feasible.

- The control is described by Theorem 3.1 for $t \geq T^u$.

- Prior to time T^u, there is no need to process any material which will not be used to erase the shortfall.

- Consider any two consecutive machines, as shown in Figure 3. If machine M_i can process material as fast as (or faster than) the maximum rate at which machine M_{i+1} processes material, then there is no need for it to transfer the contents of its buffer to buffer b_{i+1} until the exact moment that the material will be processed by machine M_{i+1} as well.

- The upstream machines in section S_i should not produce until all of the downstream buffers in section S_i have emptied, and should then produce at a rate matching the rate of machine M_{s_i}.

- Once all of the buffers (excluding b_L) downstream from any section are empty, they will remain empty.

- If b_L has a shortfall, and buffers $b_{L-1}, b_{L-2}, \ldots, b_{s_i+1}$ are empty, but section S_i contains material, then machine M_{s_i} should produce at its maximum rate μ_{s_i}.

- The first machine in each section is initially idle, and then it processes material as fast as it can, subject to its maximum processing rate, or if its buffer level is zero, at a rate matching its input, until there is no longer a shortfall.

- We call the length of time for which the first machine in each section is initially idle as the *deferral time* of that section. For convenience, let us define $l(u)$ as the index of the section that contains $M_{i(u)}$, i.e.,

$$M_{i(u)} \in S_{l(u)}. \tag{9}$$

For a given T^u, the first machine of the section that contains machine $M_{i(u)}$ should wait as long as possible before beginning to process material.

The following theorem shows that controls in $\mathcal{F}_8$ are completely characterized by the deferral times $\{\tau^u_{s_1}, \ldots, \tau^u_{s_N}\}$ of the bottleneck machines $M_{s_1}, M_{s_2}, \ldots, M_{s_N}$.

Theorem 3.2 *Consider $x_L(0) < 0$. Every $u \in \mathcal{F}_8$ has the form shown below, for some deferral times $\{\tau^u_{s_i} : 1 \leq i \leq N\}$.*

The controls $\{u_{s_i}(t) : i = 1, \ldots, N\}$ fall into three categories:
For $1 \leq i \leq l(u) - 1$,
$$u_{s_i}(t) = \begin{cases} 0 & 0 \leq t < \tau^u_{s_i} \\ d & \tau^u_{s_i} \leq t, \end{cases} \tag{10}$$

where
$$\tau^u_{s_i} := \frac{\sum_{k=s_i+1}^{L} x_k(0)}{d} \text{ for } 1 \leq i \leq l(u) - 1. \tag{11}$$

For $i = l(u)$,
$$u_{s_{l(u)}}(t) = \begin{cases} 0 & 0 \leq t < \tau^u_{s_{l(u)}} \\ \mu_{s_{l(u)}} & \tau^u_{s_{l(u)}} \leq t < T^u \\ d & T^u \leq t, \end{cases} \tag{12}$$

where T^u is given by (7).

For $l(u) < i \leq N$,

$$u_{s_i}(t) = \begin{cases} 0 & 0 \leq t < \tau^u_{s_i} \\ \mu_{s_i} & \tau^u_{s_i} \leq t < T^u_i \\ \mu_{s_{i-1}} & T^u_i \leq t < T^u_{i-1} \\ \vdots & \\ \mu_{s_{l(u)+1}} & T^u_{l(u)+2} \leq t < T^u_{l(u)+1} \\ \mu_{s_{l(u)}} & T^u_{l(u)+1} \leq t < T^u \\ d & T^u \leq t, \end{cases} \qquad (13)$$

where T^u_i is the time at which section S_i clears, under control u.

For machine $M_j \in S_i$

$$u_j(t) = \begin{cases} 0 & 0 \leq t < \tau^u_j \\ u_{s_i}(t) & \tau^u_j \leq t, \end{cases} \qquad (14)$$

where

$$\tau^u_j := \begin{cases} \tau^u_{s_i} + \frac{\sum_{k=j+1}^{s_i} x_k(0)}{\mu_{s_i}} & i(u) \leq j \leq L-1 \\ \tau^u_{s_i} + \frac{\sum_{k=j+1}^{s_i} x_k(0)}{d} & 0 \leq j < i(u). \end{cases} \qquad (15)$$

It is important to point out an extension of Theorem 3.2, which follows almost immediately from the proceeding analysis. Suppose that the demand for the product is not constant, but is time-varying and known. Assume also that

$$\sup_{t>0} d(t) \leq \mu_{s_1},$$

where, from (5), μ_{s_1} is the maximum production rate of the slowest machine(s) in the system. Then, all of the conditions for controls in $\mathcal{F}_8$ continue to hold, except the optimal control for $t \geq T^u$, which will directly follow the demand $d(t)$. Thus, an optimal control will still have the same form, i.e., characterized by deferral times. However, in order to determine the optimal control, it must be remembered that the cost of the output buffer b_L will be more tedious to compute.

Consider a vector $(i, T, T_N, T_{N-1}, \ldots, T_{l(u)+1})$ where $l(u)$ is such that $M_i \in S_{l(u)}$. The following lemma shows the necessary and sufficient conditions to be satisfied by this vector such that it corresponds to a control $u \in \mathcal{F}_8$.

Lemma 3.1 Given $(i, T, T_N, T_{N-1}, \ldots, T_{l(u)+1})$ with $l(u)$ such that $M_i \in S_{l(u)}$, there exists a control $u \in \mathcal{F}_8$ with $i(u) = i$, $T^u = T$ and $T_j^u = T_j$ for $l(u)+1 \leq j \leq N$, if and only if the following conditions are satisfied:

$$0 \leq T_N \leq T_{N-1} \leq \cdots \leq T_{l(u)+1} \leq T \tag{16}$$

$$0 \leq \tau_{s_j} \leq T_{j+1} \quad \text{for} \quad l(u) \leq j \leq N-1 \tag{17}$$

$$\mu_{s_{l(u)}}(T - \tau_{s_{l(u)}}) \leq X_{l(u)}(0) \tag{18}$$

$$\sum_{j=l(u)+1}^{N}(T_j - T_{j+1})\mu_{s_j} + (T - T_{l(u)+1})\mu_{s_{l(u)}} = -x_L(0) + Td \tag{19}$$

where $T_{N+1} := 0$,

$$X_j(0) := \sum_{k=s_{j-1}+1}^{s_j} x_k(0) \text{ for } l(u) \leq j \leq N, \tag{20}$$

$$\tau_{s_{N-1}} := \frac{T_N(\mu_{s_{N-1}} - \mu_{s_N}) + X_N(0)}{\mu_{s_{N-1}}}, \tag{21}$$

and

$$\tau_{s_j} := \frac{T_{j+1}(\mu_{s_j} - \mu_{s_{j+1}}) + \tau_{s_{j+1}}\mu_{s_{j+1}} + X_{j+1}(0)}{\mu_{s_j}} \text{ for } l(u) \leq j \leq N-2. \tag{22}$$

We now calculate the cost of a control $u \in \mathcal{F}_8$ parameterized by $(i, T, T_N, T_{N-1}, \ldots, T_{l(u)})$.

Lemma 3.2 *Consider the control $u \in \mathcal{F}_8$ parameterized by $(i, T, T_N, T_{N-1}, \ldots, T_{l(u)})$, with $M_i \in S_{l(u)}$, satisfying (16)-(22). Denote by*

$$V(i, T, T_N, \ldots, T_{l(u)}) := J(u)$$

the cost of the control. Then

$$V(i, T, T_N, \ldots, T_{l(u)}) = \sum_{j=1}^{L} k_j, \tag{23}$$

where

$$k_j = \frac{c_j x_j(0)}{d}\left[\sum_{k=j}^{L} x_k(0) - \frac{1}{2}x_j(0)\right] \quad \text{for} \quad 0 \leq j < i \quad (24)$$

$$= \frac{c_i}{2}\left[(\mu_{s_{l(u)}} - d)T^2 - \mu_{s_{l(u)}}\tau_i^2 + \frac{\left(\sum_{k=i}^{L} x_k(0)\right)^2}{d}\right] \quad \text{for} \quad j = i \quad (25)$$

$$= \frac{c_j}{2}\left[\mu_{s_{n-1}}\tau_{s_{n-1}}^2 - \mu_{s_n}\tau_j^2 + (\mu_{s_n} - \mu_{s_{n-1}})T_n^2\right] \text{for } i < j \leq L-1, j = s_{n-1}+1 \quad (26)$$

$$= c_j\left[x_j(0)\tau_j + \frac{1}{2}\frac{x_j^2(0)}{\mu_{s_n}}\right] \quad \text{for } i < j \leq L-1,\ M_j \in S_n,\ j \neq s_{n-1}+1 \quad (27)$$

$$= \frac{c_{\bar{L}}}{2}\left[T^2(\mu_{s_{l(u)}} - d) + \sum_{k=l(u)+1}^{N} T_k^2(\mu_{s_k} - \mu_{s_{k-1}})\right] \quad \text{for } j = L. \quad (28)$$

$$(T - \tau_{s_{l(u)}})\mu_{s_{l(u)}} + \sum_{j=l(u)+1}^{N} X_j(0) = -x_L(0) + T d \quad (29)$$

Using the properties obeyed by a control in $\mathcal{F}_8$, we can determine lower and upper bounds on how many machines will be used to eliminate the shortfall at the output buffer and on the time required to do this. Let $x_L(0) < 0$. There exists an optimal control such that

$$T_{\min} \leq T^u \leq T_{\max} \quad \text{and} \quad i_{\max} \geq i(u) \geq i_{\min},$$

where

$$T_{\min} = -\max_{i_{\max} \leq k \leq L-1} \frac{\sum_{j=k+1}^{L} x_j(0)}{\bar{\mu}_k - d}, \quad (30)$$

$$T_{\max} = \frac{\sum_{j=i_{\min}+1}^{L-1}[\frac{\bar{\mu}_{i_{\min}}}{\bar{\mu}_j} - 1]x_j(0) - x_L(0)}{\bar{\mu}_{i_{\min}} - d}, \quad (31)$$

$$i_{\max} := \max\{0 \leq i \leq L-1 : \sum_{j=i}^{L-1} x_j(0) \geq -x_L(0) + d \max_{i \leq k \leq L-1} \frac{\sum_{j=i}^{k} x_j(0)}{\bar{\mu}_k}\}, \quad (32)$$

and

$$i_{\min} := \max\{0 \leq i \leq L-1 : \sum_{j=i}^{L-1} x_j(0) \geq -x_L(0) + d \sum_{j=i}^{L-1} \frac{x_j(0)}{\bar{\mu}_j}\}, \quad (33)$$

where $\bar{\mu}_j$ is defined in (6).

The following theorem now shows how the optimal control problem is determined by solving $(i_{max} - i_{min} + 1)$ quadratic programming problems.

Theorem 3.3 *For each $i \in \{i_{min}, i_{min}+1, \ldots, i_{max}\}$, let $V_{min}(i)$ be the optimal cost of the quadratic program:*

$$\text{Minimize } V(i, T, T_N, \ldots, T_{l(u)+1}) \tag{34}$$

subject to (16)-(22) where $l(u)$ is such that $M_i \in S_{l(u)}$. Let $(T^{(i)}, T_N^{(i)}, \ldots, T_{l(u)+1}^{(i)})$ be the optimizing solution. Also let

$$V^* = \min_{i_{min} \leq i \leq i_{max}} V_{min}(i) \tag{35}$$

be the minimum of the costs over i, and let i^* be the minimizing choice of i. Then V^* is the optimal cost of the control problem, and the control in $\mathcal{F}_8$ corresponding to $(i^*, T^{(i^*)}, T_N^{(i^*)}, \ldots, T_{l(u)+1}^{(i^*)})$ is an optimal control.

Perkins and Kumar [31] determined the optimal control for a two-machine example. As a by-product of this example, it was shown that for the parameter values considered, it is not optimal to erase the initial shortfall in minimum time.

We generalize this example, by explicitly calculating the optimal control for the two-machine system, regardless of initial conditions and system parameters. As will be shown, the optimal control is dependent on linear switching surfaces. Using Theorem 3.3, this form of the optimal control can be generalized to L machine systems.

Example 3.1 *Determination of the optimal control for a general two-machine system.*

Consider the general two-machine system shown in Figure 4. If $\mu_0 \geq \mu_1$, then there is only one section. Thus, the resulting optimal control is trivial, i.e., there are no deferral times. The more interesting case occurs if $\mu_0 < \mu_1$. If $x_2(0) \geq 0$, then the optimal control is again trivial, and is given by Theorem 3.1. If $x_2(0) < 0$, then the optimal value for the deferral time τ_0 must be calculated.

Using Theorem 3.3, the optimal control depends on the value of $x_1(0) + \alpha x_2(0)$. Explicitly,

$$u_0(t) = \begin{cases} 0 & t \leq \tau_0 \\ \mu_0 & \tau_0 \leq t \leq T \\ d & t \geq T \end{cases}$$

where T is time at which the shortfall is eliminated.

This relationship is illustrated in Figure 5. It is important to note that, for the two-machine system with $\mu_0 < \mu_1$ and $x_2(0) < 0$, if the initial conditions line in Region 3, then the optimal control will <u>never</u> be a minimum time control (nor will it be a maximum time control). This underscores the importance of explicitly including buffer costs in the problem formulation.

Regardless of the initial conditions, the deferral time τ_0 can easily be computed directly from Figure 5. Also, the effect of an increase (or decrease) is capacity can be trivially evaluated, e.g., if $\mu_0 \geq \mu_1$, there is no benefit in investing in additional capacity for machine M_0.

The constants in Figure 5 are given by

$$\alpha = \frac{c_2^-(\mu_1 - \mu_0)}{c_1^-(\mu_0 - d) + c_2^-(\mu_1 - d)} \quad \text{and} \quad \beta = \frac{\mu_1}{\mu_1 - d}.$$

Since $\dot{x}_1 = u_0(t) - u_1(t)$ and $\dot{x}_2 = u_1(t) - d$, the optimal control in each region is determined by the buffer levels flows. In Region 1, since there is a surplus in the output buffer, the optimal control allows the output buffer to drain until it is empty, i.e.,

$$\dot{x}_1 = 0 \quad \text{and} \quad \dot{x}_2 = -d.$$

In Region 2, there is enough material in buffer b_1 to erase the shortfall. The optimal control in this region is given by

$$\dot{x}_1 = -\mu_1 \quad \text{and} \quad \dot{x}_2 = \mu_1 - d.$$

The optimal control in Region 3 is the same as in Region 2. The difference is that in Region 3 there is not enough material to erase the shortfall. (Thus, the cost-to-go functions differ in form for the two regions.) The boundary between Regions 2 and 3 is described by the equation $x_1 = -\alpha x_2$.

Note that, once in Region 3, the system will necessarily reach Region 4. The line dividing Region 3 and Region 4, where $x_1 = -\alpha\, x_2$, determines the deferral time for machine M_0. In Region 3, machine M_0 is idle; whereas, in Region 4, the machine is working at maximum capacity, i.e.,

$$\dot{x}_1 = \mu_0 - \mu_1 \quad \text{and} \quad \dot{x}_2 = \mu_1 - d.$$

The description of the optimal control is completed by the determination of the control on the axes. As discussed previously, the optimal control for $x_2 \geq 0$ is given by Theorem 3.1. On the negative x_2-axis, buffer b_1 is empty, thus the optimal control is to pipeline material, at rate μ_0, from machine M_0 directly to the output buffer. □

This simple example illustrates several important characteristics of the optimal control, which remain valid for manufacturing systems with an arbitrary number of machines. For a multiple machine system, the optimal control is again determined by a series of linear switching hyperplanes. In general, these can be calculated off-line (dependent only on the order relationship between the machine processing rates) using Theorem 3.3. Then, the optimal control is determined by a threshold comparison of linear combinations of the buffer levels.

By the Principle of Optimality, we can arbitrarily set time equal to 0. Thus, for systems with perturbations such as machine failures, we can operate under the optimal deterministic control, once the machines are functional. It is important to note that it is not necessary to re-calculate the optimal switching curves for each initial condition, i.e., the curves will require re-evaluation only when there is a change in capacity in the system.

Perkins and Kumar [31] also determined the optimal control for re-entrant lines with nondecreasing buffer costs for the case in which the final machine is the bottleneck. It is hoped that the derivation will provide insight into the general problem, where the processing times at the machines are arbitrary. This problem remains open, and will not be discussed further here.

Next, we will examine the case in which there is machine unreliability.

4 Buffer Level Control of Unreliable Pull Manufacturing Systems

The second subproblem focuses on the optimal scheduling of n part-types in a failure-prone single machine manufacturing system. This type of problem can be thought of as arising when a single machine is isolated in the re-entrant system, and it is buffered using regulators to smooth burstiness in the material flow (see Cruz [14, 15], Humes [20] and Perkins, Humes, and Kumar [28]).

This subproblem uses the previously described model, but also assumes that the machine is subject to occasional breakdowns. The machine is failure-prone with the time between failures (or the "up" time) and the time between repairs (or the "down" time) being exponentially distributed with means $1/q_d$ and $1/q_u$, respectively.

The dynamics of $x_i(t)$, the buffer level of part-type i, satisfy the following differential equation:

$$\frac{dx_i}{dt} = u_i(t) - d_i, \quad i = 1, ..., n, \tag{36}$$

where d_i is the constant demand rate for part-type i and $u_i(t)$ is the production rate of part-type i. For all $t \geq 0$, the production rates are constrained by

$$\sum_{i=1}^{n} u_i(t)\tau_i \leq 1, \tag{37}$$

where τ_i is the time required to produce one unit of part-type i. In order to simplify the expressions involved in obtaining the optimal solutions, we will assume that $\tau_i = 1/\mu$ for every part-type i, where μ is the maximum production rate for all parts. This condition can be easily relaxed without any conceptual or mathematical difficulty.

The scheduling problem is to determine $u(t) = [u_1(t), \ldots, u_n(t)]^T$ such that the average cost

$$J = \lim_{T \to \infty} \frac{1}{T} E(\int_0^T \sum_{i=1}^{n} [c_i^-(x_i^-) + c_i^+(x_i^+)]dt) \tag{38}$$

is minimized. As before, $x_i^+ = \max(0, x_i)$ and $x_i^- = \max(0, -x_i)$. The information available to the production schedule module at any time instant t is the buffer levels $x(s)$, $0 \leq s \leq t$.

4.1 LSC Control for One Machine, Two Part-Types

Quadratic approximations to the differential cost-to-go function, which yield linear switching curves, have been extensively studied in the literature. Indeed, the results of the first part of this paper prove that linear switching curve controls are optimal for systems with reliable machines. With this motivation, we provide solutions to the partial differential equations associated with the steady-state probability density functions of the buffer levels for two part-type, single machine flexible manufacturing systems under a linear switching curve (LSC) policy. When there are more than two part types, we derive the density functions under a prioritized hedging point (PHP) policy by decomposing the multiple part type problem into a sequence of two part-type problems. The expressions for the steady-state probability density functions are independent of the cost function. Therefore, for additive cost functions, we can compute the optimal PHP policy, or, in the case of only two part-types, the more general optimal LSC policy.

Consider the single machine, two part-type system shown in Figure 6, which is a special case of the system presented in Section 2. Let $x_j(t)$ be the level of buffer b_j at time t. Thus, $x_j(t) \in (-\infty, \infty)$, where a negative buffer level indicates a shortfall. Buffer b_j is continuously depleted at a constant rate d_j, indicating a demand for part-type j. Assume that there is an infinite surplus of raw material for each part-type, which is allowed to flow into the system at rate $u_j(t)$, where $u_1(t) + u_2(t) \leq \mu$.

Suppose that a linear switching curve (LSC) control is implemented by the machine. That is, when the machine is up, the control attempts to regulate the buffer levels so that they lie on the line defined by

$$(\mu_2 - d_2)(z_1 - x_1) = (\mu_1 - d_1)(z_2 - x_2),$$

and they do not exceed the hedging point (z_1, z_2). Specifically, when the machine is up,

$$(u_1, u_2) = \begin{cases} (0,0) & \text{if } x_1 > z_1, x_2 > z_2 \\ (d_1, d_2) & \text{if } x_1 = z_1, x_2 = z_2 \\ (\mu, 0) & \text{if } (\mu_2 - d_2)(z_1 - x_1) < (\mu_1 - d_1)(z_2 - x_2) \\ (\mu_1, \mu_2) & \text{if } (\mu_2 - d_2)(z_1 - x_1) = (\mu_1 - d_1)(z_2 - x_2) \\ (0, \mu) & \text{if } (\mu_2 - d_2)(z_1 - x_1) > (\mu_1 - d_1)(z_2 - x_2). \end{cases}$$

The LSC control is parameterized by the choice of the hedging points z_1 and z_2 and the control μ_1, where $\mu_2 := \mu - \mu_1$. Without loss of generality, we will assume that

$$0 \leq \mu_1 < \frac{d_1}{d_1 + d_2} \mu. \tag{39}$$

Note that, if we set $\mu_1 = d_1$, then the resulting control is a prioritized hedging point (PHP) scheduling policy.[1]

The following lemma can be shown along the lines of Srivatsan [37].

Lemma 4.1 *If the capacity condition*

$$q_u \mu - (d_1 + d_2)(q_u + q_d) > 0 \tag{40}$$

is satisfied, then, under the LSC control,

(i) the system is stable, i.e.,

$$\lim_{t \to \infty} \frac{1}{t} \sum_{i=1,2} E(x_i^2(t)) < \infty,$$

(ii) the non-transient region in the $x_1 \times x_2$ space is the set of points (x_1, x_2) that satisfy

$$(\mu_2 - d_2)(z_1 - x_1) \leq (\mu_1 - d_1)(z_2 - x_2),$$

and

$$(z_1 - x_1) d_2 \geq (z_2 - x_2) d_1,$$

and (iii) the system is ergodic.

---

[1]That is, part-type 1 has priority over part-type 2, and the machine will feed buffer b_1 at rate μ as long as $x_1 < z_1$.

The direction and magnitude of material flow in the non-transient region (in $x_1 \times x_2$ space) is shown in Figure 7. The flows are depicted separately for the "up" (State 1) and "down" (State 0) states. For the purposes of presentation, it is convenient to think of the steady-state density function, denoted by $m(x_1, x_2)$, as consisting of five components:

1. $f_0(x_1, x_2)$ which represents the density function in the interior of the non-transient space when the machine is in State 0.

2. $f_1(x_1, x_2)$ which represents the density function in the interior of the non-transient space when the machine is in State 1.

3. $R(x_1, x_2)$ which represents the density function in State 1 along the line defined by
$$(\mu_2 - d_2)(z_1 - x_1) = (\mu_1 - d_1)(z_2 - x_2).$$

4. $L(x_1, x_2)$ which represents the density function in State 0 along the line defined by
$$d_2(x_1 - z_1) = d_1(x_2 - z_2).$$

5. A point mass at (z_1, z_2) denoted by γ.

Partial differential equations describing the function $m(x_1, x_2)$ can be derived (see, for example, Algoet [2]) using probability flow conservation principles. However, these are extremely difficult to solve directly. In Section 4.2, we will present a linear transformation that yields partial differential equations which can be solved using a Laplace transform technique.

4.2 Linear Transformation

Consider the following linear transformation:

$$x := (\mu_2 - d_2)(z_1 - x_1) - (\mu_1 - d_1)(z_2 - x_2), \quad y := -d_2(z_1 - x_1) + d_1(z_2 - x_2). \quad (41)$$

Thus, the dynamics of $x(t)$ and $y(t)$ are given by

$$\begin{aligned}\dot{x} &= -(\mu_2 - d_2)(u_1 - d_1) + (\mu_1 - d_1)(u_2 - d_2) \\ \dot{y} &= d_2(u_1 - d_1) - d_1(u_2 - d_2).\end{aligned} \quad (42)$$

Under this transformation, the fluid flow model in terms of x and y is shown in Figure 8, where, for a given LSC control,

$$a := d_2\,\mu;\ b := \mu_2(\mu - d_1 - d_2);\ c := d := d_1\,\mu_2 - d_2\,\mu_1. \quad (43)$$

Thus, the capacity condition (40) can be restated as

$$b\,d\,q_u > c(a+d)\,q_d. \quad (44)$$

If we consider the (x,y) system in steady-state, then the probability density function can be represented as five components: two functions of two variables, $p_1(x,y)$ and $p_0(x,y)$, two functions of one variable, $w(y)$ and $s(x)$, and a point mass γ located at $(0,0)$, as shown in Figure 8.

Since we assume that the system is in steady-state, the system flows in both the x and the y directions must total to 0. Using this conservation of probability flow, it follows that

$$\gamma = \frac{b\,d\,q_u - c(a+d)\,q_d}{b\,d\,(q_d + q_u)} > 0.$$

As is shown in [33], the steady-state probability densities satisfy the following equations:

$$-b\,\frac{\partial p_1(x,y)}{\partial x} + a\,\frac{\partial p_1(x,y)}{\partial y} = -q_d\,p_1(x,y) + q_u\,p_0(x,y), \quad (45)$$

$$c\,\frac{\partial p_0(x,y)}{\partial x} = q_d\,p_1(x,y) - q_u\,p_0(x,y), \quad (46)$$

$$-\frac{c\,d}{q_d}\,\frac{dp_0(0,y)}{dy} = -c\,p_0(0,y) + b\,p_1(0,y), \quad (47)$$

$$\frac{c}{q_u}\,\frac{dp_1(x,0)}{dx} = -p_1(x,0), \quad (48)$$

where

$$w(y) = \frac{c}{q_d}\,p_0(0,y),\ s(x) = \frac{a}{q_u}\,p_1(x,0), \quad (49)$$

$$p_0(0,0) = \frac{q_d^2}{c\,d}\,\gamma,\ \text{and}\ p_1(0,0) = \frac{q_d\,q_u}{a\,c}\,\gamma. \quad (50)$$

Solving equation (48), using (49), yields

$$p_1(x,0) = \frac{q_d q_u \gamma}{a c} e^{-\frac{q_u}{c} x}. \qquad (51)$$

Substituting this into equation (46) gives

$$p_0(x,0) = \frac{q_d^2 \gamma}{c d}[1 + \frac{d q_u}{a c} x]e^{-\frac{q_u}{c} x}. \qquad (52)$$

Since $x \in [0,\infty)$ and $y \in [0,\infty)$, we can use a Laplace transform technique to solve for the steady-state probability densities. This will be shown in Section 4.3.

4.3 Steady-State Probability Densities

Taking the Laplace transform of (45)-(47) with respect to the variable y, we obtain a set of ordinary differential equations. The inverse Laplace transform of the solution to these equations yields the steady-state probability densities:

$$p_0(x,y) = e^{-(q_u x/c + \beta y)} \{-\delta_1 \frac{2b}{a} \frac{\partial}{\partial x}[I_0(\epsilon\sqrt{y^2 + \frac{a x y}{b}})] - \\ \delta_2 \frac{2b}{a} e^{-(\lambda-\beta)y} \frac{\partial}{\partial x}[\int_0^y e^{(\lambda-\beta)\tau} I_0(\epsilon\sqrt{\tau^2 + \frac{a x \tau}{b}}) d\tau] + \delta_3 I_0(\epsilon\sqrt{y^2 + \frac{a x y}{b}}) + \\ \delta_4 \frac{\partial}{\partial y}[I_0(\epsilon\sqrt{y^2 + \frac{a x y}{b}})] + \delta_5 e^{-(\lambda-\beta)y} \int_0^y e^{(\lambda-\beta)\tau} I_0(\epsilon\sqrt{\tau^2 + \frac{a x \tau}{b}}) d\tau\} \qquad (53)$$

and

$$p_1(x,y) = e^{-(q_u x/c + \beta y)} \{-\frac{2b\theta_1}{a} \frac{\partial}{\partial x}[I_0(\epsilon\sqrt{y^2 + \frac{a x y}{b}})] - \frac{2b\theta_2}{a} \frac{\partial^2}{\partial x \partial y}[I_0(\epsilon\sqrt{y^2 + \frac{a x y}{b}})] - \\ \theta_3 \frac{2b}{a} e^{-(\lambda-\beta)y} \frac{\partial}{\partial x}[\int_0^y e^{(\lambda-\beta)\tau} I_0(\epsilon\sqrt{\tau^2 + \frac{a x \tau}{b}}) d\tau] + \theta_4 I_0(\epsilon\sqrt{y^2 + \frac{a x y}{b}}) + \\ \theta_5 e^{-(\lambda-\beta)y} \int_0^y e^{(\lambda-\beta)\tau} I_0(\epsilon\sqrt{\tau^2 + \frac{a x \tau}{b}}) d\tau\}, \qquad (54)$$

where

$$\beta := \frac{b q_u + c q_d}{a c}, \quad \lambda := \frac{b q_u}{c(a+d)} - \frac{q_d}{d}, \quad \epsilon := \sqrt{\frac{4 b q_d q_u}{a^2 c}},$$

and $I_0(\cdot)$ is the modified Bessel function of order 0, i.e.,

$$I_0(x) = \sum_{k=0}^{\infty} \frac{(x/2)^{2k}}{(k!)^2}.$$

The constants in (53) and (54) are given by

$$\delta_1 = \frac{(a+2d) q_d \gamma}{2 c(a+d)}, \quad \delta_2 = \frac{d q_d \gamma}{2 c(a+d)}[\frac{(a+d) q_d}{d^2} - \frac{b q_u}{c(a+d)}], \quad \delta_3 = \frac{q_d^2 \gamma}{c d},$$

$$\delta_4 = \frac{q_d\,\gamma}{c}, \quad \delta_5 = \frac{q_d^2\,\gamma}{a\,c}\,[\frac{(a+d)\,q_d}{d^2} - \frac{b\,q_u}{c\,(a+d)}],$$

and

$$\theta_1 = -\frac{a\,d\,q_u\,\gamma}{2\,c\,(a+d)^2}, \quad \theta_2 = \frac{a\,d\,\gamma}{2\,b\,(a+d)}, \quad \theta_3 = \frac{d^2\,q_u\,\gamma}{2\,c\,(a+d)^2}\,[\frac{(a+d)\,q_d}{d^2} - \frac{b\,q_u}{c\,(a+d)}],$$

$$\theta_4 = \frac{(a+2\,d)\,q_d\,q_u\,\gamma}{a\,c\,(a+d)}, \quad \theta_5 = \frac{d\,q_d\,q_u\,\gamma}{a\,c\,(a+d)}\,[\frac{(a+d)\,q_d}{d^2} - \frac{b\,q_u}{c\,(a+d)}].$$

The following theorem provides the relationship between the steady-state density in the (x,y) system and the steady-state density in the single machine, two part-type (x_1, x_2) system.

Theorem 4.1 *The steady-state densities of the buffer levels are given by*

$$\begin{aligned} f_j(x_1, x_2) &= c\,p_j(x, y), \quad j = 0, 1, \\ L_1(x_1) &= (c/d_1)\,s(x), \\ R_1(x_1) &= \frac{c}{\mu_1 - d_1}\,w(y), \\ L_2(x_2) &= (d_1/d_2)\,L_1(x_1), \\ R_2(x_2) &= \frac{\mu_1 - d_1}{\mu_2 - d_2}\,R_1(x_1), \end{aligned}$$

and the equations (49), (53), and (54).

□

Thus, the steady-state probability density functions are known for any system which can be transformed so that it has flow diagrams of the form shown in Figure 8. It follows that, for any such system, it is straightforward to evaluate any cost function which depends only on the steady-state buffer level probability densities.

For performance criteria which are separable, such as the additive cost criterion, it is only necessary to determine the marginal probability densities for each of the two buffers. The marginal densities can be derived from the expressions in the previous section. However, the marginal probability densities can also be derived directly from the original system. In order to do this, the two-dimensional components of the joint probability density along the axes are needed:

$$p_0(0,y) = \frac{q_d \gamma}{c} e^{-\lambda y} \left[\frac{q_d}{d} - \frac{a}{2(a+d)} \int_0^y e^{-(\beta-\lambda)t} \frac{\epsilon I_1(\epsilon t)}{t} dt \right] \quad (55)$$

$$p_1(0,y) = \frac{d\gamma}{a+d} \left[\frac{a}{2b} e^{-\beta y} \frac{\epsilon I_1(\epsilon y)}{y} + \frac{q_u}{c} e^{-\lambda y} \left(\frac{q_d}{d} - \frac{a}{2(a+d)} \right. \right.$$
$$\left. \left. \times \int_0^y e^{-(\beta-\lambda)t} \frac{\epsilon I_1(\epsilon t)}{t} dt \right) \right]. \quad (56)$$

See Perkins and Srikant [33] for the details of this approach.

We now extend these results to an arbitrary number of part-types.

4.4 Steady-State Densities for Multiple Part-Types Under the PHP Policy

In this section we show that, given the expression for the steady-state joint density function for the two part-type problem under the PHP (prioritized hedging point) policy, the steady-state density for each buffer of an n part-type problem can be derived by suitably decomposing the n part-type problem into a sequence of two part-type problems.

Note that the PHP policy is a zero-inventory or Just-in-Time (JIT) policy if

$$(z_{s_1}, z_{s_2}, \ldots, z_{s_n}) = (0, 0, \ldots, 0).$$

There are several good reasons to consider PHP policies:

1. When the optimal policy is a zero-inventory or Just-in-Time (JIT) policy, then the optimal PHP policy is also JIT. Thus, optimality conditions in the class of PHP policies provide necessary conditions for the optimality of a JIT policy. This is a very important consideration given the fact that JIT policies have strongly been favored in real-life production systems for process discipline reasons even when they are not optimal.

2. If the cost functions are not linear, then the most natural extension of the hedging point policy of the single part-type case is the PHP policy.

3. Such policies have been studied in literature for more general problems (see, for example, Perkins and Kumar [30] and Cornelius [13]).

Let $\pi(x_1, x_2, d_1, d_2, \mu)$ be the joint density function of the steady state buffers for a two-part problem under the PHP policy. The last variables represent the dependence of this function on the parameters of the two part-type problem. The following theorem presents a decomposition technique by which the marginal densities of each of the buffers of an n part-type system can be obtained from the expression for the two part-type problem.

Theorem 4.2 *For $1 \leq i \leq n-1$, let $p(\tilde{x}_{s_i}, x_{s_{i+1}})$ denote the joint density of $\tilde{x}_{s_i}$ and $x_{s_{i+1}}$, where $\tilde{x}_{s_i} := \sum_{k=1}^{i} x_{s_k}$. Then, under the PHP policy, $p(\tilde{x}_{s_i}, x_{s_{i+1}})$ is given by*

$$p(\tilde{x}_{s_i}, x_{s_{i+1}}) = \pi(\tilde{x}_{s_i}, x_{s_{i+1}}, \sum_{k=1}^{i} d_{s_k}, d_{s_{i+1}}, \mu).$$

The proof is provided in Perkins and Srikant [33].

4.5 Minimization of the Average Cost Function

In this section, we present results for obtaining the optimal hedging point for a given LSC policy, when the marginal probability density for each buffer level is known. To do this, we first show the translation invariance of the steady-state buffer level cumulative distribution functions under any feedback policy, and then specialize this result to LSC policies. The translation invariance is used to derive a condition for the optimality of the hedging point as in Bielecki and Kumar [6]. Additional details and explanation can be found in Perkins and Srikant [32].

From Srivatsan [37], we know that the system is stable and ergodic under a hedging point policy, where a hedging point policy is defined as follows:

Hedging Point Policy: Any feedback policy such that, for all $t \geq 0$,

$$\sum_{i=1}^{n} u_i(t) = \begin{cases} 0 & \text{if } x_i \geq z_i \text{ for all } i \\ \sum_{i=1}^{n} d_i & \text{if } x_i = z_i \text{ for all } i \\ \mu & \text{otherwise} \end{cases}$$

where $(z_1, z_2, \ldots, z_n)$ is the hedging point.

Let $M_i(x_i, z_i)$ denote the steady-state cumulative distribution function for part-type i, $i = 1, 2, \ldots, n$ under a specific feedback policy. Now, we have the following lemma:

Lemma 4.2 *Under a hedging point policy, the steady-state cumulative distribution function for each buffer level is translation invariant with respect to the hedging point, i.e., for all i*

$$M_i(x_i, z_i) = M_i(x_i - z_i, 0).$$

For a given hedging point policy, the following theorem describes how to determine the optimal values for the hedging points.

Theorem 4.3 *Consider an n part-type, single machine system. Define $m_i(0, z_i)$ to be the steady-state marginal probability density for the buffer level of part-type i, under a given hedging point policy having hedging point $(z_1, z_2, \ldots, z_n)$. For this form of hedging point policy, the choice of z_i which minimizes the average cost function (38) is given by*

$$z_i^* = \max\left\{0, \left\{z_i : \int_{-\infty}^{0} m_i(x_i, z_i)\, dx_i = \frac{c_i^+}{c_i^- + c_i^+}\right\}\right\}. \tag{57}$$

We will give an outline of the proof, when a linear switching curve scheduling policy is implemented. For an n part-type system, under the LSC policy, the cost (38) can be expressed as

$$J_{LSC} = \sum_{i=1}^{n} J_{LSC}^i,$$

where

$$J_{LSC}^i = -\int_{-\infty}^{0} c_i^- \, x_i m_i(x_i, z_i)\, dx_i + \int_{0}^{z_i} c_i^+ \, x_i \, m_i(x_i, z_i)\, dx_i + c_i^+ \, \gamma_i \, z_i,$$

and γ_i is the point mass at z_i. Note that J_{LSC} is separable. Thus, the determination of the optimal value for each z_i can be performed independently. Using Lemma (4.2) and the fact that

$$\int_{-\infty}^{z_i} m_i(x_i)\, dx_i + \gamma_i = 1,$$

it follows that

$$\frac{\partial J_{LSC}^i}{\partial z_i} = -(c_i^- + c_i^+) \int_{-\infty}^{0} m_i(x_i, z_i)\, dx_i + c_i^+. \qquad (58)$$

□

Note that the right-hand side of (57) does not depend of the choice of z_i. Thus, the optimal value of z_i is found simply by varying z_i until the left-hand side of (57) reaches the desired value.

We have assumed that, for each part-type, the marginal probability density function exists for all $x_i < z_i$ and it has only one discontinuity at $x_i = z_i$. This is a nonrestrictive assumption that easily can be relaxed to allow a countable number of discontinuities. Specifically, the cumulative distribution function can be allowed to have point masses at a countable number of points.

5 Concluding Remarks

We have studied two subproblems based on a fluid flow model of a manufacturing system. For the deterministic problem, we have shown that an optimal control can be determined by solving a series of quadratic programming problems having linear constraints. It is desirable to extend these results to more general buffer cost structures and machine routings. It should be noted that, although the first proposed subproblem considers machines which do not fail, the results can be used to derive bounds on manufacturing systems which are nearly reliable, i.e., systems in which the machines rarely fail or fail for only a short time compared to the system dynamics.

One of the longstanding unsolved problems in the optimal control of failure-prone manufacturing systems has been the derivation of analytic, closed-form optimal solutions except in the special case of single machine, single part-type, linear cost systems. Analytic solutions not only provide an excellent insight into the solutions for a class of problems (as opposed to a specific set of parameters), but are also useful from a computational point of view by reducing the numerical calculations involved. An important step towards determining analytical solutions has been taken, by providing closed-form expressions for the buffer level steady-state probability density functions for the two part-type, single machine problem under the linear switching curve (LSC) control. A policy called the prioritized hedging point (PHP) policy has been introduced which is a natural extension of the hedging point policy for single part-type systems, which is prevalent in the literature. The results obtained are also applicable to single machine systems with more than two part-types. The determination of analytical results is a significant research development.

A possible next step is to extend these results to include machines in tandem (series). The equations which govern the steady-state probability distribution of the buffer levels are very similar to those for the single machine system discussed above. In fact, preliminary results suggest that the methods used to solve the single machine problem are applicable.

Recently, there have been several major papers written which provide methods for deriving performance bounds for systems with discrete part flow (see, for example, Bertsimas, Paschalidis, and Tsitsiklis [5], Kumar and Meyn [24], and Kumar and Kumar [23]). It would be useful for the results obtained using the approximate fluid flow model adopted in this paper to be compared to these achievable performance levels.

In closing, we reiterate that the long-term objective of the research presented is the determination of detailed scheduling policies for two idealized subproblems. The work herein is intended to assist in laying the foundation for the integration of these scheduling policies. It is hoped that, when implemented in real-world industrial applications, these scheduling algorithms can be "tuned" in an intelligent and intuitive fashion in order to achieve desired performance levels.

6 Acknowledgments

The authors are grateful to P. R. Kumar and F.-X. Wu, for their contributions to early and more recent, respectively, portions of the work reported here.

Also, the National Science Foundation has provided partial funding for this work, under RIA Grant No. ECS-9410242.

References

[1] R. Akella and P. R. Kumar, "Optimal control of production rate in a failure prone manufacturing system," *IEEE Transactions on Automatic Control*, vol. 31, pp. 116–126, February 1986.

[2] P. Algoet, "Flow balance equations for the steady-state distribution of a flexible manufacturing system," *IEEE Transactions on Automatic Control*, vol. 34, pp. 917–921, July 1989.

[3] F. Avram, D. Bertsimas, and M. Ricard, "Fluid models of sequencing problems in open queueing networks: An optimal control approach." Working Paper. Northeastern University and MIT, September 1994.

[4] R. Bellman, *Dynamic Programming*. Princeton, NJ: Princeton University Press, 1957.

[5] D. Bertsimas, I.C. Paschalidis, and J. N. Tsitsiklis, "Optimization of multiclass queueing networks: Polyhedral and nonlinear characterizations of achievable performance," *Annals of Applied Probability*, vol. 4, pp. 43-75, 1994.

[6] T. Bielecki and P. R. Kumar, "Optimality of zero-inventory policies for unreliable manufacturing systems," *Operations Research*, vol. 36, pp. 532–541, July-August 1988.

[7] E. K. Boukas and A. Haurie, "Manufacturing flow control and preventive maintenance: A stochastic control approach," *IEEE Transactions on Automatic Control*, vol. 35, pp. 1024–1031, 1990.

[8] M. Caramanis and G. Liberopoulos, "Perturbation analysis of flexible manufacturing system flow controllers," *Operations Research*, vol. 40, pp. 1107–1125, November–December 1992.

[9] M. Caramanis and A. Sharifnia, "Near optimal manufacturing flow controller design," *International Journal of Flexible Manufacturing Systems*, vol. 3, pp. 321–336, 1991.

[10] H. Chen and A. Mandelbaum, "Discrete flow networks: Bottleneck analysis and fluid approximations," *Mathematics of Operations Research*, vol. 16, pp. 408–446, 1991.

[11] H. Chen and D. D. Yao, "Dynamic scheduling of a multiclass fluid network," *Operations Research*, vol. 41, pp. 1104–1115, November–December 1993.

[12] S. Connolly, "A real-time policy for performing setup changes in a manufacturing system," M.S. thesis, Operations Research Center, Massachusetts Institute of Technology, 1992.

[13] A. Cornelius, "An iterative simulation method for determining the optimal buffer level regulation policy for unreliable manufacturing systems," M.S. thesis, University of Illinois, Urbana, IL, 1992.

[14] R. L. Cruz, "A calculus for network delay, part I: Network elements in isolation," *IEEE Transactions on Information Theory*, vol. 37, pp. 114–131, January 1991.

[15] R. L. Cruz, "A calculus for network delay, part II: Network analysis," *IEEE Transactions on Information Theory*, vol. 37, pp. 132–141, January 1991.

[16] S. Gershwin, R. Akella and Y. Choong, "Short term production scheduling of an automated manufacturing facility," *IBM Journal of Research and Development*, vol. 29, pp. 392–400, 1985.

[17] P. Glasserman, "Hedging–point production control with multiple failure modes," *IEEE Transactions on Automatic Control*, vol. 40, pp. 707-712, April 1995.

[18] B. Hajek and R. G. Ogier, "Optimal dynamic routing in communication networks with continuous traffic," *Networks*, vol. 14, pp. 457–487, 1984.

[19] J. M. Harrison and L. M. Wein, "Scheduling networks of queues: Heavy traffic analysis of a two-station closed network," *Operations Research*, vol. 38, pp. 1052–1064, November-December 1990.

[20] C. Humes, Jr., "A regulator stabilization technique: Kumar-Seidman revisited," *IEEE Transactions on Automatic Control*, vol. 39, pp. 191–196, January 1994.

[21] J. Kimemia and S. Gershwin, "An algorithm for the computer control of production in flexible manufacturing systems," *IIE Transactions*, vol. 15, no. 4, pp. 353–362, December 1983.

[22] P. R. Kumar, "Re–entrant lines," *Queueing Systems: Theory and Applications: Special Issue on Queueing Networks*, vol. 13, pp. 87–110, May 1993.

[23] S. Kumar and P. R. Kumar, "Performance Bounds for Queueing Networks and Scheduling Policies," *IEEE Transactions on Automatic Control*, vol. 39, pp. 1600–1611, August 1994.

[24] P. R. Kumar and S. P. Meyn, "Stability of queueing networks and scheduling policies," *IEEE Transactions on Automatic Control*, vol. 40, pp. 251-260, February 1995.

[25] S. H. Lu and P. R. Kumar, "Control policies for scheduling of semiconductor manufacturing plants," *Proceedings of the 1992 American Control Conference*, Chicago, Illinois, pp. 828–829, June 1992.

[26] S. H. Lu and P. R. Kumar, "Fluctuation smoothing scheduling policies for queueing systems." To appear in *Proceedings of the Silver Jubilee Workshop on Computing and Intelligent Systems*, Indian Institute of Science, Bangalore, December 1993.

[27] J. R. Perkins, "Stable, real-time scheduling of a single machine manufacturing system," M.S. thesis, University of Illinois, Urbana, IL, 1990. Also, CSL Rept. DC-121, UILU-ENG-90-2226, UIUC.

[28] J. R. Perkins, C. Humes, Jr., and P. R. Kumar, "Distributed control of flexible manufacturing systems: Stability and performance," *IEEE Transactions on Robotics and Automation*, vol. 10, pp. 133–141, April 1994.

[29] J. R. Perkins and P. R. Kumar, "Stable distributed real-time scheduling of flexible manufacturing/assembly/disassembly systems," *IEEE Transactions on Automatic Control*, vol. 34, pp. 139–148, February 1989.

[30] J. R. Perkins and P. R. Kumar, "Buffer level control of unreliable manufacturing systems," *Proceedings of the 1992 American Control Conference*, Chicago, Illinois, pp. 826–827, June 1992.

[31] J. R. Perkins and P. R. Kumar, "Optimal control of pull manufacturing systems." To appear in *IEEE Transactions on Automatic Control*, 1995.

[32] J. R. Perkins and R. Srikant, "Hedging point policies for failure-prone manufacturing systems: Optimality of JIT policies and bounds on buffer levels." To appear in *Proceedings of the 34th IEEE Conference on Decision and Control*, 1995.

[33] J. R. Perkins and R. Srikant, "Scheduling multiple part-types in an unreliable single machine manufacturing system." Submitted to *IEEE Transactions on Automatic Control*, 1995.

[34] R. Rishel, "Dynamic programming and minimum principles for systems with jump Markov disturbances," *SIAM Journal of Control*, vol. 13, no. 2, pp. 338–371, February 1975.

[35] S. P. Sethi and X. Y. Zhou, "Feedback controls in two machine flowshops Part I: Optimal controls in the deterministic case." Submitted to *IEEE Transactions on Automatic Control*, 1994.

[36] A. Sharifnia, "Production control of a manufacturing system with multiple machine states," *IEEE Transactions on Automatic Control*, vol. 33, pp. 600–626, July 1988.

[37] N. Srivatsan, "Synthesis of optimal policies for stochastic manufacturing systems, " Ph.D. thesis, Massachusetts Institute of Technology, Cambridge, MA, 1993. Also, Laboratory for Manufacturing and Productivity Rept. LMP-93-009, MIT.

[38] J. Tsitsiklis, "Convexity of characterization of optimal policies in a dynamic routing problem," *Journal of Optimization Theory and Its Applications*, vol. 44, no. 1, pp. 105–135, September 1984.

[39] L. Wein, "Scheduling semiconductor wafer fabrication," *IEEE Transactions on Semiconductor Manufacturing*, vol. 1, pp. 115–130, August 1988.

[40] L. Wein, "Scheduling networks of queues: Heavy traffic analysis of a two-station network with controllable inputs," *Operations Research*, vol. 38, pp. 1065–1078, November-December 1990.

[41] L. Wein, "Scheduling networks of queues: Heavy traffic analysis of a multistation network with controllable inputs," *Operations Research*, vol. 40, pp. 312–334, May-June 1992.

[42] F.-X. Wu and J. R. Perkins, "Optimal control of deterministic manufacturing systems with holding costs." Working Paper. Department of Manufacturing Engineering, Boston University, 1995.

JAMES R. PERKINS
DEPARTMENT OF MANUFACTURING ENGINEERING
BOSTON UNIVERSITY
44 CUMMINGTON ST.
BOSTON, MA 02215

R. SRIKANT
COORDINATED SCIENCE LABORATORY
UNIVERSITY OF ILLINOIS
1308 W. MAIN ST.
URBANA, IL 61801

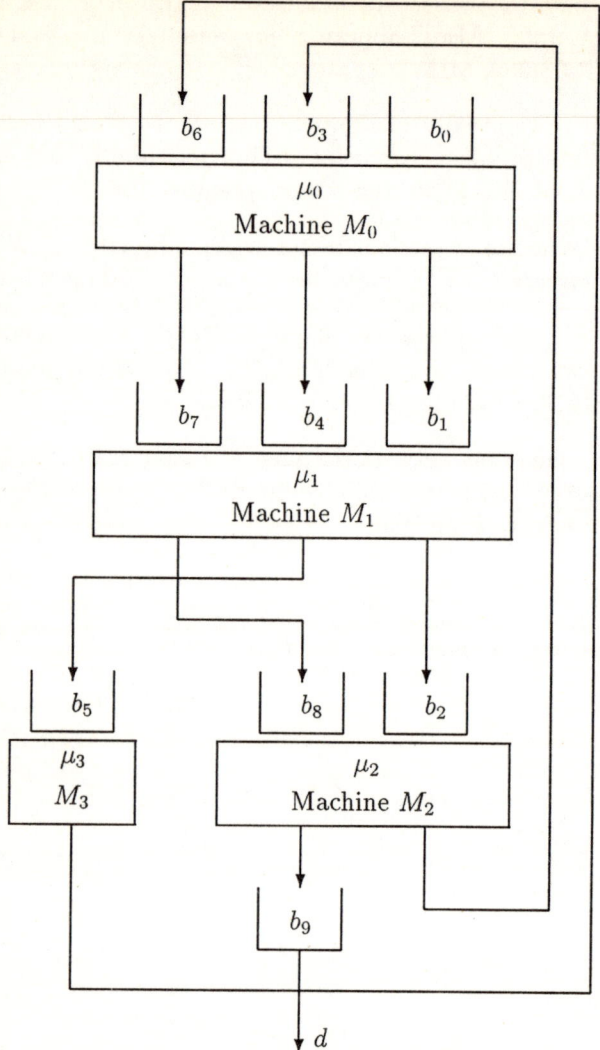

Figure 1: A re-entrant line.

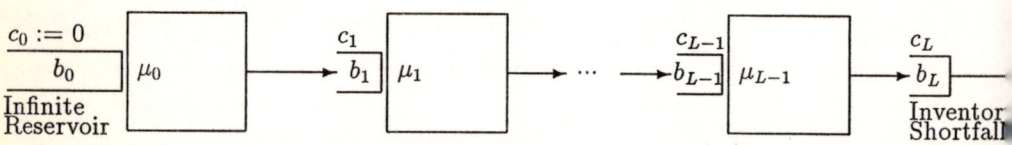

Figure 2: A flow line with external input and drainage.

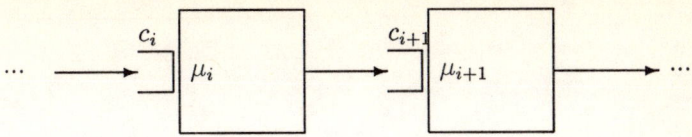

Figure 3: Two consecutive machines.

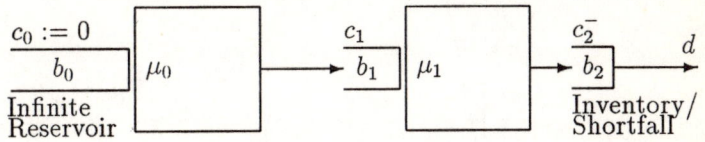

Figure 4: A two-machine system.

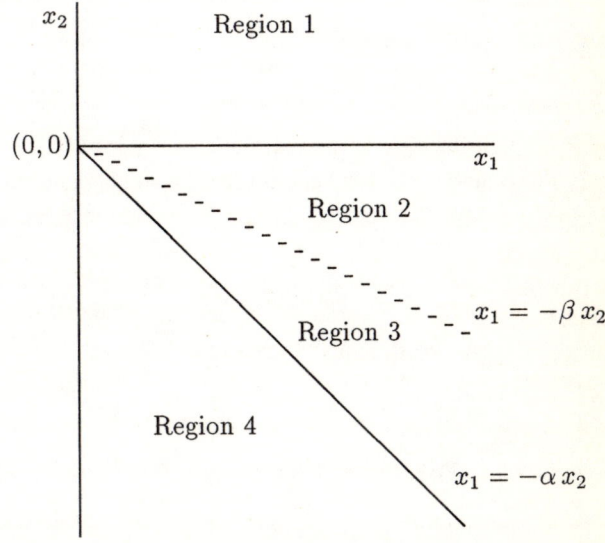

Figure 5: Optimal control regions for two-machine system.

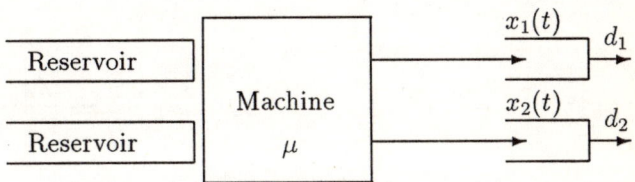

Figure 6: The single machine, two part-type system.

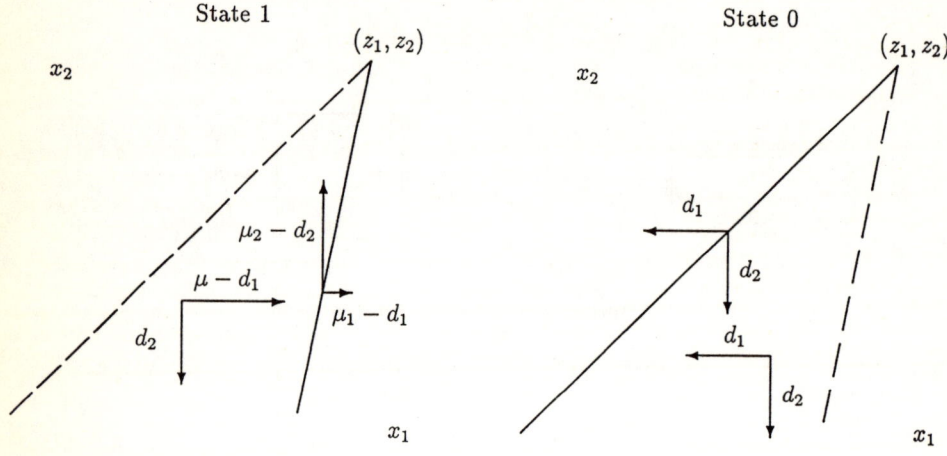

Figure 7: Flows under the LSC control for the single machine, two part-type case.

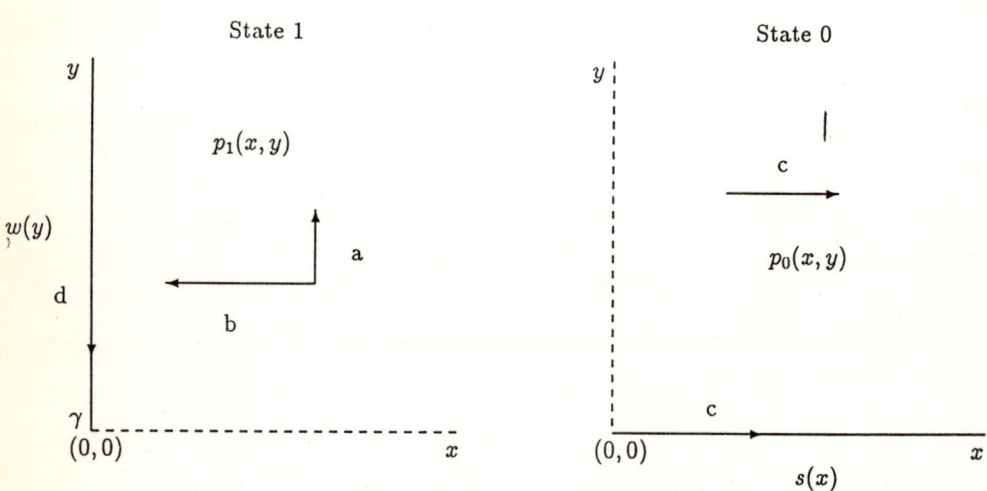

Figure 8: Flows and densities for (x, y) system.

PART II
SCHEDULING AND IMPROVABILITY

Using Second Moment Information in Stochastic Scheduling

Marilyn J. Maddox and John R. Birge[*]

Abstract

We present a scheduling heuristic for the stochastic job shop. In a stochastic job shop, operation processing times may vary randomly, or machines may fail at random intervals, or both. Dispatching rules based on substituting expected values for random quantities are often used for scheduling in this uncertain environment. The scheduling heuristic we propose performs restricted dynamic updating of an initial schedule for the shop, using limited distribution information about the random quantities. In particular, we allow general dependence among the stochastic quantities and require at most two moments of each distribution function. Our objective is to minimize expected tardiness. We investigate the benefits of the scheduling heuristic versus a dispatching rule using a simulation study.

Key words: scheduling, stochastic programs, simulation

Introduction

In the static job shop problem, n jobs must be scheduled for processing on m machines. Each job consists of a number of operations; each operation runs on a specific machine or group of machines. The order of machines a job must visit is known. In the deterministic job shop problem, machines are continuously available and processing times for all operations are deterministic. In stochastic versions of the problem, operation processing times may be random or the machines may be subject to random disruptions or both. We consider a

---

[*]The research of this author was supported in part by Office of Naval Research Grant N00014-86-K-0628 and the National Science Foundation under Grants ECS-8815101 and DDM-9215921.

stochastic version of the job shop problem. We assume that preemption of one operation by another is not allowed. In addition, we assume that setup times are negligible in comparison to operation processing times and that material handling time between operations is insignificant. Basically, ignoring our consideration of stochastic elements, these assumptions define a simple job shop as described in Conway et. al. (1967).

We concentrate on minimizing the expected tardiness in a job shop. We chose to minimize tardiness because reducing tardiness is a goal in many actual job shops. Smith et. al. (1986) report that tardiness is the single most important objective to practitioners today. In addition, since tardiness costs are nonstationary in time, results for tardiness may be extended to more general functions than results for stationary objectives such as completion time.

A schedule that is optimal with respect to minimizing expected tardiness may perform quite poorly given the actual situation in the job shop. Ideally, we desire a schedule that performs well in a variety of situations. One option is to select a schedule that does well in the worst case scenario. This minimizes the maximum cost of any scheduling problem. However, it does not insure good average performance. Nevertheless, in many cases, managing the worst case scenarios is important. In this paper, we describe a heuristic scheduling procedure which combines the average and worst-case concepts by selecting schedules which minimize over the set of possible schedules the maximum expected schedule cost over distributions with known limited information.

This paper is organized as follows. The remainder of this section describes the stochastic job shop problem and reviews the relevant literature. The second section describes the heuristic scheduling algorithm. The third section presents a simulation study which compares the expected tardiness of schedules developed using the heuristic with schedules developed under a dispatching rule. The section concludes with observations from the study and indicates directions for future research.

1 Background

1.1 Stochastic Job Shop Problem

Under the scenario of random processing times or machine disruptions, the stochastic job shop problem can be modeled as a stochastic program with general recourse. This means the optimal schedule may vary with the realized outcomes of the random components. In this model, the decision of which job to schedule next can depend on any decisions made and events that have occurred in the past, i.e., only nonanticipative decisions are allowed. Thus, the sequence of operations actually processed on a machine depends on the events which occurred during processing. This flexibility may make the general recourse schedule extremely complicated to implement. Simplification is possible by restricting consideration to simple recourse solutions, although the solution obtained is no longer guaranteed to be optimal.

In simple recourse, decisions are independent of all but the initial conditions. A simple recourse schedule for the job shop is completely specified by the processing order on each machine. Simple recourse schedules are inflexible because they do not permit scheduling changes once processing begins. In reality, it is often desirable to update the schedule periodically during processing, using current information on the job shop status. Our heuristic performs limited updating of an initial schedule based on current shop status.

Although most job shops are stochastic, scheduling models usually ignore uncertainty (Graves (1981)). In these models, expected values are substituted for their actual random counterparts. Research on stochastic problems in various applications shows that this procedure of substituting expected values yields disappointing results, see Igelmund (1983b), Kallberg et.al. (1982), Mittenthal (1986), and Wets (1983). In practice, the schedules produced by these deterministic models must be adapted to constantly changing circumstances. In very uncertain environments, a priori schedules are abandoned altogether. Instead, each time a machine becomes idle, a dynamic priority rule is used to choose one operation to process next among the operations queued at the machine. The priority rule assigns an index, based on local operation information, to each operation in the queue. Priority rule scheduling ignores future implications of current decisions. In addition, it reduces the possibility of closely coordinating a job shop with its associated manufacturing processes.

The heuristic described here creates an initial schedule for the job shop before production begins. As processing proceeds, this schedule is updated based on current information on the job shop status. Thus, the heuristic incorporates the strengths of both simple recourse scheduling and dynamic priority rule scheduling. It uses up-to-date information about the job shop status and analyzes the effects of current decisions on the future schedule.

1.2 Literature Review

Much of the existing literature on the stochastic job shop problem is devoted to special cases, either restricting the number or type of machines or requiring explicit distributions for the random elements. The processing times may be stochastic or the machines may be subject to random breakdowns or both.

Many of the stochastic scheduling articles focus on the single machine problem. Both the stochastic job and stochastic machine cases are represented. Some early work on a stochastic job problem was done by Blau (1973). More recently, Glazebrook (1983, 1984, 1987a, 1987b) considered both stochastic job and stochastic machine scenarios. Mittenthal (1986), Mittenthal and Birge (1987), and Birge et. al. (1990) used stochastic programming to model the stochastic single machine problem. Most of these papers require complete knowledge of the distribution function of the random elements. Independence among random elements is also a common assumption.

There is little published work on the multi-machine, stochastic job shop problem. Assuming independent, exponential job processing times, Weiss and Pinedo (1980) found rules for processing n jobs on nonidentical processors to minimize stationary objectives.

Igelmund and Radermacher (1983a, 1983b) and Möhring, Radermacher and Weiss (1984, 1985) studied a stochastic, resource constrained, project scheduling problem. They assumed that the durations of project activities, equivalent to operations in job shop scheduling, were random. They also assumed that the joint distribution of the durations of the activities was known. They considered simple recourse solutions as well as schedules created from a combination of several simple recourse solutions. The authors proved various analytical properties for these classes of schedules.

Solel (1986) presented a general precedence constrained stochastic scheduling model which encompasses the models of Möhring et. al. and Weiss and Pinedo. She modeled the problem as a Piecewise Deterministic (Markov) process and, assuming that the joint distribution of the job durations was known, obtained existence results and characterized the optimal strategy when the objective was to minimize the expectation of a cost function. The cost functions she considered had to be linearly bounded and monotonically increasing as a function of the completion times of the jobs.

All of the models discussed so far explicitly take uncertainty into account. Job shop rescheduling is a heuristic for the stochastic problem that ignores randomness during the planning stage. In job shop rescheduling, a schedule for the job shop is initially constructed using deterministic methods. When a disruption occurs, the job shop is rescheduled from that point onward. The basic dilemma in this heuristic is how to reschedule quickly enough that work flow in the job shop is not disturbed. Of issue are the frequency of rescheduling, whether the whole job shop should be rescheduled, and which deterministic scheduling heuristic should be used. Rescheduling has been studied by Muhleman et. al. (1982), Bean et. al. (1991), Bean and Birge (1986), and Birge (1986), among others.

The stochastic scheduling problem we study is more general than most of the models discussed in the stochastic scheduling literature. We consider multiple machines and assume only limited information about the distributions of the random elements. In particular, we allow general dependence among the stochastic quantities and require at most the first two moments of each distribution function. Due to the computational difficulty of solving general recourse stochastic programs, and the complexity of the resulting solutions, we consider a heuristic solution procedure in this paper.

2 Heuristic Scheduling Procedure

The first step in the heuristic algorithm is to create an initial schedule for the job shop prior to actual processing. This schedule dictates the first operation to load on each machine. The remaining steps of the scheduling heuristic control sequencing once production begins. Each time an operation finishes processing, the heuristic updates the current schedule and selects an operation to put on the machine next. The heuristic stops when all operations have finished processing.

This section consists of two subsections. The first subsection describes a method for creating an initial schedule for the job shop. The second subsection discusses a method for updating this schedule as production progresses.

2.1 Initial Schedule

The heuristic algorithm begins by creating an initial schedule for the job shop. In realistic scheduling environments, initial schedules which balance several objectives may be desired. Here, we schedule with the objective of minimizing expected tardiness. The initial schedule may be created by substituting expected values for random quantities and using a deterministic scheduling method. For example, a deterministic dispatching rule may be used. On a small problem, a branch and bound algorithm discussed in Maddox (1988) may be used to create an initial schedule which has the minimum worst-case expected tardiness among all such schedules. Evaluating the true expected tardiness of the initial schedule is difficult because of the stochastic nature of the problem and the number of possible alternatives.

2.2 Updating the Schedule

When an operation finishes processing, the heuristic updating procedure determines the next operation to place on that machine. The heuristic chooses either the operation specified in the current schedule or selects a replacement operation. Since the procedure sequentially updates the initial schedule, there is always a complete schedule available. This permits evaluation of the future effects of current decisions. If the updating is too extensive, the heuristic essentially becomes a dispatching rule, since the future sequencing information it utilizes is unreliable. If the information the algorithm uses is incorrect, the algorithm could potentially create schedules with higher tardiness than an algorithm which ignores such information. On the other hand, not allowing enough revisions may make the heuristic algorithm too inflexible, given the stochastic nature of the job shop. When an operation finishes processing, the key question is what operation to put on the machine next. Hence, in our heuristic algorithm we only permit updates of the next operation to be placed on the machine.

We refer to the machine involved in the current decision as the current machine, M. Any machines identical to the current machine are identified as the current machine group, N. M is included in machine group N. The process of deciding which operation to load on machine M involves four steps:

- Identify the set of potential "replacement operations" for operation I, the operation scheduled next on machine M in the current schedule.

- For each replacement operation, develop a replacement schedule in which the replacement operation is scheduled next on machine M.

- Choose the "best" replacement operation from the set of potential replacements.

- Load either operation I or the "best" replacement operation on machine M. The first three steps are described in more detail below.

Step 1: Identify Potential Replacement Operations

The number of potential replacement operations affects both the speed and the effectiveness of the algorithm. If a large number of potential replacements are identified, evaluating which one is "best" may require a significant amount of time. On the other hand, limiting the number of replacements limits the flexibility of the algorithm.

The set of potential replacements could include all unprocessed operations on machine group N. However, unless the current job shop conditions are drastically different from the expected ones, operations currently sequenced "farther" from the current time T are less likely to be good replacements than operations scheduled closer to T. Technological precedence constraints are also likely to prohibit moving operations not currently scheduled "near" T.

Of particular interest as replacements are operations that are now scheduled to start before or soon after the expected completion time of operation I in the current schedule. When the variance of operation I increases, the set of replacement operations should increase as well. Therefore, we included an operation in the set of replacement operations if it ran on machine group N, and it was expected to start before the expected completion time of operation I plus a constant times the standard deviation of operation I's processing time. The constant is a parameter of the algorithm.

Step 2: Develop Replacement Schedules

Once an operation has been identified as a potential replacement, the heuristic develops a schedule with that operation placed next on machine M. Let J denote a replacement operation. If J is originally scheduled on machine M the operations scheduled between operations I and J are bumped back one position in the replacement schedule. If operation J is scheduled on another machine in machine group N, the two operations are interchanged in the replacement schedule.

The positions of the predecessor and successor operations of operation J should also be updated in the replacement schedule. This increases the impact of the schedule change. The predecessor and successor operations must be resequenced carefully, since incorrectly placing them may cause unwanted delays and machine idle time. Determining their new positions is difficult, however, since the exact completion times of the unprocessed operations are unknown.

In the heuristic algorithm, if replacement operation J is now sequenced before an operation H which it was originally scheduled after, the algorithm checks to see if any successors of H are scheduled before successors of J on a machine. If any such successors are found, interchanges are made to rectify the situation. In the simulation, a version of the heuristic with no successor updates is tested as well, to judge the viability of this approach.

Step 3: Choose the "Best" Replacement Operation

Ideally, once each potential replacement schedule has been developed, the schedule with the lowest expected tardiness is selected as the replacement schedule. If the expected tardiness of this replacement schedule is less than that of the current schedule, the replacement operation is loaded on the current machine and the replacement schedule becomes the current schedule. Unfortunately, the expected tardiness of a schedule cannot be calculated exactly. The tardiness random variable has a complicated distribution since it is a function of the maximum of a sum of possibly dependent random variables. It is possible, however, to bound the expected tardiness of a schedule. Birge and Maddox (1995) and Maddox (1988) derive an upper bound on the expected tardiness of a schedule. A lower bound on the expected tardiness of a schedule can be obtained by substituting expected values for the random processing times and determining the resulting tardiness.

These bounds bracket expected tardiness in some interval. We use this interval to determine the operation to schedule. Using the lower bound assumes a "best-case" distribution. The upper bound assumes a "worst-case" distribution. The schedule with either the minimum best-case or worst-case expected tardiness becomes the current schedule. The operation dictated by this schedule is loaded onto machine M.

Developing and bounding the expected tardiness of each replacement schedule may be time consuming. An alternative is to choose one replacement operation and build only one replacement schedule. This requires a procedure for determining the "best" replacement from the set of potential replacement operations. The goal is to determine the operation which is most likely to have the schedule with the smallest bound on expected tardiness. One possible way to pick the operation is to use a dispatching rule. Dispatching rules which perform well in studies minimizing tardiness are most appropriate. In the version of the heuristic tested in the simulation, a replacement schedule is built for each replacement operation.

3 Simulation

3.1 Simulation Description

The purpose of the simulation was to compare the average tardiness performance of the heuristic algorithm and a dynamic dispatching rule. Dynamic dispatching rules are often used to schedule stochastic, multiple machine job shops. The heuristic algorithm tested was described in the previous section. The initial schedule was generated assuming operation processing times equaled their expected values. The Apparent Tardiness Cost (ATC) dispatching rule (Vepsalainen and Morton (1987)), described below, was used to generate the initial schedule. Each time an operation completed processing, a set of potential replacement operations was identified and two replacement schedules were developed for each. In one replacement schedule, only the position of the potential replacement operation was altered. In the other, the position of the replacement operation and the positions of all of its technological successors were updated. The length of the time interval in which replacements are chosen is a parameter of the simulation. The length is a function of the standard deviation of operation I's processing time.

Under each set of experimental conditions, two forms of the heuristic were tested. In one, the schedule with the minimum worst-case expected tardiness, among the replacement schedules and the current schedule, was chosen as the new current schedule. The bounding procedure discussed in Birge and Maddox (1995) and Maddox (1988), utilizing two moments of each processing time distribution, was used to determine the value of each schedule. In the other version of the heuristic, the schedule with the minimum best-case expected tardiness was chosen. Note that both of these approaches recognize the stochastic nature of the problem but make different assumptions about the distribution of processing times. The choice of best-case versus worst-case depends on the decision-maker's risk aversion. The ATC dispatching rule was chosen for testing against the heuristic because of its superior performance in deterministic studies. The ATC of operation j of job i is

$$\frac{1}{E(\xi_j)} exp(-[\frac{T_i - t - \bar{\xi}_j - \sum_{q=j+1}^{l_i}(W_q + \bar{\xi}_q)}{kp}]^+),$$

where

t = current time,

W_q = expected wait time for remaining operation q,

l_i = number of last operation in job i,

p = expected average processing time of waiting operations,

k = lookahead parameter.

The expected wait time for operation q is set equal to $b \cdot \bar{\xi}_q$, where b is a parameter of the simulation. In our simulation, $b = 2$ and $k = 3$. Vepsalainen and Morton (1987) suggest $k = 3$ as a reasonable value for dynamic job shops.

Here, the expected processing time of an operation was used in calculating its ATC. Initially, the ATC of each operation without a predecessor was calculated. The one with maximum ATC was loaded on machine M. This process continued by selecting the operation with maximum ATC from the set of operations whose predecessors had been scheduled, until all operations completed processing.

The job shop selected for study is part of an automobile manufacturing facility consisting of 14 machine groups and 26 machines. Ten machine groups contain a single machine, two groups consist of three machines, one machine group contains four machines and the last machine group includes six machines. The machines are all subject to random breakdowns, with a mean time of 2400 minutes between breakdowns. The time to repair each machine is 60 minutes. Means and variances for operation processing times are calculated from this machine information. Fifteen jobs require scheduling on these machines. In total, ninety-three operations must be scheduled. The due dates for the jobs are based on total work content.

The ATC dispatching rule and the two versions of the heuristic were simulated under four parameter combinations of due date tightness and variance in the operation processing times. The due dates were set to 2 or 2.21 times the expected processing times. When the due dates were set to 2 times the expected processing times, there were no replications with zero tardiness. When the due dates were set to 2.21 times the expected processing times, the ATC dispatching rule produced schedules with zero tardiness in two thirds of the replications. Two variance levels were tested in the simulation. In one case, the variance in the processing times was calculated assuming 2400 minutes between breakdowns and 60 minute repair times. In the other case, the variance in the processing times was calculated with double or triple the repair times.

3.2 Simulation Results

The simulation consisted of three experiments. Each experiment used a different distribution to generate the operation processing times. The three distributions tested were: a lognormal distribution with independent processing times, a bi-modal distribution with independent processing times, and a bi-modal distribution with totally correlated processing times.

The results of the simulation study using lognormal operation processing times appear in Table 1 .The ATC dispatching rule is identified by ATC in the table. In Table 1, BC,c=x refers to the version of the heuristic which chooses the schedule with the minimum best-case expected tardiness at each decision point. Replacements are chosen from the interval of length equal to the expected processing time plus c times the standard deviation of operation I. WC,c=x refers to the version of the heuristic which chooses the schedule with the minimum worst-case tardiness at each decision point. Table 1 lists the median and mean sample tardiness for each scheduling procedure under each parameter combination. Under each parameter combination, the scheduling procedures are entered in order of median tardiness. The table also lists the confidence level that the probability of a negative difference between the tardiness of schedules created by adjacent scheduling procedures is greater than the probability of a positive difference. It includes, for each parameter combination, the number of replications (out of 15 total for each combination) in which a schedule created by the WC,c=x or ATC procedures has lower tardiness than the corresponding schedule created by the BC,c=2 algorithm. The final column in Table 1 summarizes the tardiness range, rounded to the nearest integer, of the schedules obtained under each parameter combination.

Initially, using tight due dates and low variance, the WC version of the heuristic was run with different c values to judge the effect of c on the results. Table 1 shows that the median and mean tardiness values were similar under the various choices of c. Setting c=10 gives the worst results for the WC algorithm. Increasing the number of potential replacements, while allowing additional flexibility in the algorithm, does not necessarily produce schedules with lower tardiness, because inaccurate future information is being used in decision making. Since the version of the algorithm with c=2 performed best in these tests, the remainder of the simulation runs were completed with c=2.

Note that ranking the scheduling algorithms by sample mean expected tardiness produces a different order. This is primarily because of an extremely large expected tardiness value, an order of magnitude greater than any other, on one of the replications. The ATC dispatching rule always produced the

Parameter Combination	Scheduling Procedure	Median Tardiness	Mean Tardiness	Confidence Level	# Better Than BC	Tardiness Range
tight due date low variance	BC, c=2	27.277	200.26	99	-	22-2256
	WC, c=2	54.485	222.26	50	1	22-2261
	WC, c=0	54.485	354.12	60	2	22-4256
	WC, c=10	57.510	353.94	50	1	50-4109
	ATC	76.291	213.86	-	1	23-1861
tight due date high variance	BC, c=2	25.971	412.71	99	-	21-5346
	WC, c=2	54.893	443.15	50	0	24-5351
	ATC	78.173	367.83	-	1	51-4021
loose due date low variance	BC, c=2	0.000	129.1	99	-	0-1729
	WC, c=2	8.818	136.03	60	0	0-1730
	ATC	16.011	110.06	-	1	0-1285
loose due date high variance	BC, c=2	0.000	339.37	99	-	0-4818
	WC, c=2	7.897	348.2	50	0	0-4820
	ATC	14.030	261.19	-	1	0-3446

Table 1: Simulation Results with Lognormal Processing Times

schedule with the lowest tardiness on this case. Removing this replication, the order of the means is the same as the order of the medians.

Altering the due date tightness and variance levels does not influence the ranking of the scheduling procedures. Under every parameter combination with lognormal processing times, the BC version of the heuristic produces schedules with the lowest median expected tardiness. In every case, there is 99algorithm is more likely to produce lower expected tardiness schedules than the other algorithms. Comparing the WC,c=2 and ATC procedures, the highest confidence that the probability of a negative difference in tardiness is greater than the probability of a positive difference in tardiness occurs when the due dates are tight and the variance is low.

In Maddox (1988), Chapter 4, for two test problems, the simple recourse schedule with minimum best-case expected tardiness was tested against the simple recourse schedule with minimum worst-case tardiness. There, when a lognormal distribution was used to generate the processing times, it was concluded that the worst-case schedule performed well only in extreme situations. In light of this, it is not surprising that the BC version of the heuristic performs better in these tests, in terms of average tardiness, than the WC version of the heuristic. Given the small number of replications in the current study, statistics on the behavior of the algorithms in extreme situations is inappropriate. The cost of greatly increasing the number of replications in this study was prohibitive. We conclude, however, that the behavior of the BC and WC heuristic algorithms is similar to the behavior of the optimal simple recourse schedules.

The use of lognormal processing times is not likely to be representative of a shop in which machines generally perform in a fixed amount of time unless a tool breaks or another disruption occurs. In this case, the distribution of each operation processing time is more likely to follow a bi-modal distribution. The distribution achieving the worst-case expected tardiness is also a bi-modal distribution (see Birge and Maddox, 1995). For this reason, bi-modal distri-

Parameter Combination	Scheduling Procedure	Median Tardiness	Mean Tardiness	# Best (of 30)	Tardiness Range
tight due date low variance	BC, c=2	120	168.4	0.5	30-240
	WC, c=2	60	85.2	24	120-540
	ATC	120	118.1	4.5	30-240
tight due date high variance	BC, c=2	330	479.1	0.5	120-2190
	WC, c=2	60	301.0	17.5	30-1080
	ATC	120	225.5	11.5	30-750
loose due date low variance	BC, c=2	0	18.8	12	0-216
	WC, c=2	18	33.9	0	0-234
	ATC	0	10.7	18	0-156
loose due date high variance	BC, c=2	0	168.2	10.83	0-1215
	WC, c=2	18	159.7	1.33	0-1146
	ATC	0	92.2	17.83	0-474

Table 2: Simulation Results for Bi-Modal, Independent Processing Times

butions were also tested. The first set of results considers only independent processing times. This represents the situation in which processing is only loosely connected among machines. It is not the worst-case distribution because the worst-case distribution has many correlated processing times. The results of this experiment appear in Table 2. In these results, many of the values were the same because the set of possible processing times was limited. Instead of calculating confidence levels for achieving lower tardiness, we simply list the number of times out of 30 replications in which each method achieved the lowest overall tardiness. Ties were split equally among the tying methods.

Note the difference in the results of Table 2 compared to Table 1. Here, tight due dates favor the WC rule with ATC again producing lower mean values when variance is high. For looser due dates, ATC is preferred. This observation is consistent with the small problem results in Maddox (1988). Tight due dates tend to favor the forward looking but pessimistic WC rule over the dynamic ATC rule and the optimistic BC rule. This characteristic appears even when the processing times are independent.

The next set of results considers completely correlated processing times. This situation may occur, for example, when a batch of parts or its pallet has a defect that delays all processes. It is also representative of a shop of tightly coupled machines, where a breakdown in one area quickly shuts down other areas of the shop. Completely correlated processing times are not necessarily the worst-case distribution because that distribution may impose anti-correlations in order to achieve higher overall expectations. It is probably the worst-case in most practical situations.

The results for this experiment appear in Table 3. Here, there are three possible scenarios (corresponding to no disruptions, only long jobs disrupted and all jobs disrupted). In this case, the WC heuristic obtains minimum expected tardiness when due dates are tight. ATC performs well in the worst outcomes although in the middle outcomes (only long jobs disrupted) ATC always yielded the highest tardiness.

The computational results demonstrate that the heuristic algorithm with limited updating is a valuable scheduling method in some scenarios. The worst-case algorithm appears preferable when processing times are bi-modal and particularly when processing times are correlated. The best-case algorithm appears best in expectation in the uni-modal cases. The dynamic scheduling approach, as represented by ATC, appears best when processing time variation is high. These results are consistent with tests on small models in Maddox (1988).

Two alternative versions of the heuristic were tested in this simulation.

Parameter Combination	Scheduling Procedure	Median Tardiness	Mean Tardiness	# Best (of 3)	Tardiness Range
tight due date low variance	BC,c=2	120	2804	0	120-5469
	WC, c=2	60	2503	3	60-4930
	ATC	120	2673	0	120-5046
tight due date high variance	BC, c=2	120	12139	0	120-22371
	WC, c=2	60	11360	3	60-20784
	ATC	120	11491	0	120-20877
loose due date low variance	BC, c=2	0	2399	1.5	0-5232
	WC, c=2	18	2094	0	0-4269
	ATC	0	2128	1.5	0-4179
loose due date high variance	BC, c=2	0	11840	0.5	0-23091
	WC, c=2	18	10847	1.0	0-20244
	ATC	0	10872	1.5	0-20007

Table 3: Simulation Results for Bi-Modal, Correlated Processing Times

The difference between them was the criterion used to pick the "best" schedule to use each time an operation finished processing. Both versions could be improved by resequencing the predecessors of a potential replacement operation in the replacement schedule. In the current versions of the heuristic, the schedule for each potential replacement operation is developed and compared with the current schedule. An alternative is to pick a single replacement, develop the corresponding replacement schedule and compare it with the current schedule. This would decrease the computation time of the heuristic significantly but it could also decrease the performance of the algorithm. The relative value of these changes depends on the specific implementation.

Additional tests that examine the effect of using different objective functions for choosing the current schedule would also be valuable. One possibility is to use the probability that tardiness is greater than some threshold value as the objective. This objective may more accurately reflect the costs of tardiness as premium freight charges are incurred. Initial bounds for this quantity appear in Maddox and Birge (1991).

Acknowledgments

The authors gratefully acknowledge the assistance of Sharon Fox in performing the computational tests.

References

Bean, J.C., and J.R. Birge (1986), "Match-up Real-time Scheduling," in R.H. F. Jackson and A. W. T. Jones, eds., *Real-Time Optimization in Automated Manufacturing Facilities*, NBS Special Publication 724, National Bureau of Standards.

Bean, J.C., J.R. Birge, C.E. Noon, and J. Mittenthal (1991), "Match-up Scheduling with Multiple Resources, Release Dates and Disruptions," Operations Research 39, 470-483.

Birge, J.R. (1986), "Real-Time Adaptive Scheduling in Flexible Manufacturing Systems," in J. White, ed., *Proceedings of Material Handling Focus '85*,

Springer-Verlag, 1986.

Birge, J.R., J.B.G. Frenk, J. Mittenthal, and A.H.G. Rinnooy Kan (1990), "Single Machine Scheduling Subject to Machine Breakdowns," Naval Research Logistics Journal 37, 661-677.

Birge, J.R., and M.J. Maddox (1995), "Bounds on Expected Project Tardiness," Operations Research 43, 838-850.

Birge, J.R., and R.J-B. Wets (1986), "Designing Approximation Schemes for Stochastic Optimization Problems, in particular, Stochastic Programs with Recourse," Mathematical Programming Study, 27, 54-102.

Blau, R.A. (1973), "N-Job, One Machine Sequencing Problems Under Uncertainty," Management Science, 20, 1, 101-109.

Conway, R.W., W.L. Maxwell, and L.W. Miller (1967), *Theory of Scheduling*, Addison-Wesley Publishing Company, Inc.

Glazebrook, K.D. (1983), "On Stochastic Scheduling Problems with Due Dates," International Journal of Systems Science, 14, 11, 1259-1271.

Glazebrook, K.D. (1984), "Scheduling Stochastic Jobs on a Single Machine Subject to Breakdowns," Naval Research Logistics Quarterly, 31, 251-264.

Glazebrook, K.D. (1987a), "Sensitivity Analysis for Stochastic Scheduling Problems," Mathematics of Operations Research, 12, 2, 205-223.

Glazebrook, K.D. (1987b), "Evaluating the Effects of Machine Breakdowns in Stochastic Scheduling Problems", Naval Research Logistics, 34, 319-335.

Graves, S. C. (1981), "A Review of Production Scheduling," Operations Research, 29, 646-675.

Kallberg, J., R. White, and W. Ziemba (1982), "Short Term Financial Planning Under Uncertainty," Management Science, 28, 670-682.

Igelmund, G., and F.J. Radermacher (1983a), "Preselective Strategies for the Optimization of Stochastic Project Networks Under Resource Constraints," Networks, 13, 1-28.

Igelmund, G., and F.J. Radermacher (1983b), "Algorithmic Approaches to Preselective Strategies for Stochastic Scheduling Problems," Networks, 13, 29-48.

Maddox, M.J. (1988), "Scheduling a Stochastic Job Shop to Minimize Tardiness Objectives," Ph.D. Dissertation, The University of Michigan, Ann Arbor, Michigan.

Maddox, M.J., and J.R. Birge (1991), "Bounds on the Distribution of Tardiness in a PERT Network", Working Paper, Department of Industrial and Operations Engineering, University of Michigan.

Malcolm, D.G., J.H. Roseboom, C.E. Clark, and W. Fazar (1959), "Application of a Technique for Research and Development Program Evaluation," Operations Research, 7, 5, 646-669.

Mittenthal, J. (1986), "Single Machine Scheduling Subject to Random Breakdowns," Ph. D. Dissertation, The University of Michigan, Ann Arbor, Michigan.

Mittenthal, J., and J.R. Birge (1987), "Recourse Bounds in Stochastic Single Machine Scheduling," Technical Report #37-87-133, Decision Sciences and Engineering Systems Department, Rensselaer Polytechnic Institute.

Möhring, R.H., F.J. Radermacher, and G. Weiss (1984), "Stochastic Scheduling Problems I - General Strategies," Zeitschrift für Operations Research, 28, 7A, 193-260.

Möhring, R.H., F.J. Radermacher, and G. Weiss (1985), "Stochastic Scheduling Strategies II - Set Strategies," Zeitschrift für Operations Research, 29, 3A, 65-104.

Muhlemann, A.P., A.G. Lockett, and C.K. Farn (1982), "Job Shop Scheduling Heuristics and Frequency of Scheduling," International Journal of Production Research, 20, 227-241.

Smith, M.L., R. Ramesh, R.A. Dudek, and E.L. Blair (1986), "Characteristics of U.S. Flexible Manufacturing Systems - A Survey," in *Proceedings of the Second ORSA/TIMS Conference on Flexible Manufacturing Systems: Operations Research Models and Applications*, K.E. Stecke and R. Suri, eds., Elsevier Science Publishers B.V., Amsterdam, 477-486.

Solel, E. (1986), "A Dynamic Approach to Stochastic Scheduling via Stochastic Control," Ph.D. Dissertation, Dalhousie University, Halifax, Nova Scotia.

Vepsalainen, A.P.J., and T.E. Morton (1987), "Priority Rules for Job Shops with Weighted Tardiness Costs,"Priority Rules for Job Shops with Weighted Tardiness Costs," Management Science 33, 1035-1047.

Weiss, G. and M. Pinedo (1980), "Scheduling Tasks with Exponential Service Times on Nonidentical Processors to Minimize Makespan or Flow Time," Journal of Applied Probability, 17, 187-202.

Wets, R. J-B (1983), "Stochastic Programming: Solution Techniques and Approximation Schemes," in *Mathematical Programming: The State of the Art 1982*, A. Bachem, M. Grötschel, and B. Korte, eds., Springer-Verlag, Berlin, 566-603.

Marilyn J. Maddox
Ford Motor Co.
Information Systems Planning
Suite 425
555 Republic Drive
Allen Park, MI 48101

John R. Birge
Industrial and Operations
Engineering Department
University of Michigan
1205 Beal Ave.
Ann Arbor, MI 48109

Improvability of Serial Production Lines: Theory and Applications

D. Jacobs, C.-T. Kuo, J.-T. Lim, and S.M. Meerkov

Abstract

A production system is called improvable if limited resources involved in its operation can be redistributed in such a manner that a performance index is improved. The production systems considered in this work are the serial machining and assembly lines with unreliable machines and finite buffers. The limited resources involved are the total work-in-process (WIP) and the workforce (WF). The performance index addressed is the throughput. It is shown that the system is unimprovable with respect to the WF if and the only if each buffer is on the average half full. The system is unimprovable with respect to WIP and WF simultaneously if and only if the above holds and, in addition, each machine has equal probability of blockage and starvation. Thus, in a well designed system each buffer should be mostly half full and each machine should be starved and blocked with equal frequency.

The improvability indicators, mentioned above, have been applied to a machining line at an automotive component plant. Both WF and WIP improvability measures have been considered. It was shown that the throughput of the line can be increased by a factor of 2.

Key words: Manufacturing, Production systems, Improvability.

1. Introduction

The term production system is used to denote a set of machines and material handling devices arranged so as to produce a desired part (or a complete product) at the output of the last machine. As such, production systems are at the heart of manufacturing. No manufacturing enterprise can be successful without an efficient production system. Good product and product design, by themselves, do not guarantee a commercial success; superior production systems are necessary. For example, three most successful new commercial products of the last 25 years – FAX, CD, and VCR– all have been invented, developed, and initially designed in the West; to-day practically all are produced in the East. More efficient production systems made it possible for the Far East manufacturers to win in the competition and capture the market.

Today, production systems are designed using: (i) past experience; (ii) computer simulation; (iii) general considerations based on manufacturing gurus teaching (Deming, Taguchi, Goldratt, etc. [1]-[3]). This makes production systems design different from design methodologies in other engineering disciplines. For example, no one designs airspace vehicles or machine tools using just (i)-(iii); these are designed using first principles and other fundamental laws pertaining to a particular engineering field; (i)-(iii) may be used but only in conjunction with the fundamental engineering knowledge available.

As a result of this status quo, production systems often operate inefficiently. In practice, it is not unusual that a production system produces at a rate far below the slowest of its machines in isolation. The main reason for this is the machine breakdowns. A breakdown of a machine leads to the starvation of the downstream machine and to the blockage of the upstream machine. The material handling system, through its buffer, is supposed to attenuate these perturbations. However, when the knowledge of the laws that govern the behavior (statics and/or dynamics) of the production systems is missing, design of an efficient material handling system becomes a matter of intuition and luck. Quite often, one runs out of luck, and as a result, a production system, capable of producing, say, 10 parts/min, produces at a rate of 5 parts/min, due to the blockages and starvations.

The work described in this paper is intended to contribute to the development of engineering knowledge for production systems. We introduce and analyze a system-theoretic property of this manufacturing: improvability. Roughly speaking, a production system is improvable if limited resources, such as work-in-process (WIP) or workforce (WF), can be redistributed among its operations so that a performance index is improved. We derive quantitative conditions, observable on the factory floor, that show whether the system is improvable or not. We also report an application of these improvability indicators to a machining line at an automotive component plant.

The outline of the paper is as follows: In Section 2, the theory of improvability is presented. Section 3 describes the application. In Section 4, the conclusions are formulated. Many details and all proofs are omitted from this publication; they can be found in [4].

2. Theory

2.1. Problem Formulation

Consider a production system of M unreliable machines and B finite buffers interconnected by a material handling system. Assume that each machine is characterized by its average production rate in isolation, p_i, $i = 1, ..., M$, and each buffer is characterized by its capacity, N_i, $i = 1, ..., B$. Assume that the N_i's and p_i's are constrained as follows:

$$\sum_{i=1}^{B} N_i = N^*, \qquad (2.1)$$

$$\prod_{i=1}^{M} p_i = p^*. \qquad (2.2)$$

Constraint (2.1) implies that the total work-in-process (WIP) available in the system cannot exceed N^*. Constraint (2.2) is interpreted as a bound on the workforce (WF). Indeed, in many systems, assignment of the workforce (both machine operators and skilled trades for repair and maintenance) defines the production rate and the average up-time of each machine. Therefore, the total WF available can be conceptually mapped into constraint (2.2).

Let $PI(p_1, ..., p_M, N_1, ..., N_B)$ be the performance index of interest. Examples of PI are the average production rate, the due-time performance, product quality, etc.

Definition 2.1: A production system is called **improvable with respect to WIP** if there exists a sequence $N_1^*, ..., N_B^*$ such that $\sum_{i=1}^{B} N_i^* = N^*$

and
$$PI(p_1, ..., p_M, N_1^*, ..., N_B^*) > PI(p_1, ..., p_M, N_1, ..., N_B).$$

Definition 2.2: A production system is called **improvable with respect to WF** if there exists a sequence $p_1^*, ..., p_M^*$ such that $\prod_{i=1}^M p_i^* = p^*$ and
$$PI(p_1^*, ..., p_M^*, N_1, ..., N_B) > PI(p_1, ..., p_M, N_1, ..., N_B).$$

Definition 2.3: A production system is called **improvable with respect to WIP & WF simultaneously** if there exist sequences $N_1^*, ..., N_B^*$ and $p_1^*, ..., p_M^*$ such that $\sum_{i=1}^B N_i^* = N^*$, $\prod_{i=1}^M p_i^* = p^*$, and
$$PI(p_1^*, ..., p_M^*, N_1^*, ..., N_B^*) > PI(p_1, ..., p_M, N_1, ..., N_B).$$

Problem 2.1: Given a production system described above, find both quantitative and qualitative indicators of improvability with respect to WIP, WF, and WIP & WF simultaneously.

When a system is unimprovable under (2.1) and (2.2), constraint relaxation (i.e. increase p^* or N^*) is necessary to improve PI. The question arises: Which p_i and/or N_i should be increased so that the most benefits are obtained? To formulate this question precisely, introduce

Definition 2.4: Machine i is the **bottleneck machine** if
$$\frac{\partial PI(p_1, ..., p_M, N_1, ..., N_B)}{\partial p_i} > \frac{\partial PI(p_1, ..., p_M, N_1, ..., N_B)}{\partial p_j}, \forall j \neq i.$$

Definition 2.5: Buffer i is the **bottleneck buffer** if
$$PI(p_1, ..., p_M, N_1, ..., N_i + 1, ..., N_B)$$
$$> PI(p_1, ..., p_M, N_1, ..., N_j + 1, ..., N_B), \forall j \neq i.$$

Contrary to the popular belief, the machine with the smallest production rate and the buffer with the smallest capacity are not necessarily the bottleneck machine and the bottleneck buffer, respectively. In some cases, the most productive machine, i.e. the machine with the largest p_i, is the bottleneck. This happens because the inequalities in Definitions 2.4 and 2.5 depend not on a particular machine or buffer, but rather on the system as a whole. Therefore, the problem arises:

Problem 2.2: Given a production system defined by machines $p_1, ..., p_M$ and buffers $N_1, ..., N_B$ interconnected by a material handling system, identify the bottleneck machine and the bottleneck buffer.

Finally, in some cases, it is important to determine the total work-in-process N^*, such that all $N > N^*$ give practically no increase in PI. To formulate this property, introduce

$$PI(N) = \max_{\substack{N_1, ..., N_B \\ \sum_{i=1}^{B} N_i = N}} PI(p_1, ..., p_M, N_1, ..., N_B).$$

Definition 2.6: The total WIP, N^*, is said to be ϵ-**adapted** to $p_1, ..., p_M$ if

$$\frac{|PI(N^*) - PI(N)|}{PI(N^*)} < \epsilon, \quad \forall N \geq N^*.$$

Problem 2.3: Given a production system and $\epsilon > 0$, find the smallest N^* which is ϵ-adapted to the p_i's.

2.2. Improvability under Constraints

2.2.1. Model

Although Problems 2.1–2.3 are of interest in a large variety of manufacturing situations, for the purposes of this study we analyze the simplest, but archetypical, production system – the serial production line. A number of models for such lines are available in [5]-[8]. The following model is considered throughout this work:

(i) The system consists of M machines arranged serially, and $M - 1$ buffers separating each consecutive pair of machines.

(ii) The machines have identical cycle time T_c. The time axis is slotted with the slot duration T_c. Machines begin operating at the beginning of each time slot.

(iii) Each buffer is characterized by its capacity, N_i, $1 \le i \le M-1$.

(iv) Machine i is starved during a time slot if buffer $i-1$ is empty at the beginning of the time slot. Machine 1 is never starved.

(v) Machine i is blocked during a time slot if buffer i has N_i parts at the beginning of the time slot and machine $i+1$ fails to take a part during the time slot. Machine M is never blocked.

(vi) Machine i, being neither blocked nor starved during a time slot, produces a part with probability p_i and fails to do so with probability $q_i = 1 - p_i$. Parameter p_i is referred to as the **production rate** of machine i in isolation.

2.2.2. Preliminaries

Consider the following recursive procedure:

$$\begin{aligned}
p_i^b(s+1) &= p_i[1 - Q(p_{i+1}^b(s+1), p_i^f(s), N_i)], \quad 1 \le i \le M-1, \\
p_i^f(s+1) &= p_i[1 - Q(p_{i-1}^f(s+1), p_i^b(s+1), N_{i-1})], \quad 2 \le i \le M, \quad (2.3) \\
p_1^f(s) &= p_1, \qquad p_M^b(s) = p_M, \\
s &= 0, 1, 2, 3, ...,
\end{aligned}$$

with initial conditions

$$p_i^f(0) = p_i, \quad i = 1, ..., M,$$

where

$$Q(x, y, N) = \begin{cases} \frac{(1-x)(1-\alpha)}{1 - \frac{x}{y}\alpha^N} & x \ne y \\ \frac{1-x}{N+1-x} & x = y \end{cases} \quad (2.4)$$

and

$$\alpha = \frac{x(1-y)}{y(1-x)}.$$

Lemma 2.1 *The recursive procedure (2.3) is convergent, so that the limits*

$$p_i^f = \lim_{s \to \infty} p_i^f(s),$$
$$p_i^b = \lim_{s \to \infty} p_i^b(s), \qquad (2.5)$$
$$i = 1, ..., M$$

exist.

Introduce the production rate estimate as follows:

$$PR_{est} = p_M^f.$$

Theorem 2.1

$$|PR - PR_{est}| \leq \mathcal{O}(\delta), \quad \delta \ll 1.$$

2.2.3. Improvability with respect to WF

Theorem 2.2 *Serial production line (i)-(vi) is unimprovable with respect to WF if and only if*

$$p_i^f = p_{i+1}^b, \quad i = 1, ..., M-1. \qquad (2.6)$$

Corollary 2.1 *If condition (2.6) is satisfied,*

(a) *each machine i is blocked with almost the same frequency as machine $i+1$ is starved, i.e.*

$$|b_i - s_{i+1}| \sim \mathcal{O}(\delta) \quad i = 1, ..., M-1,$$
$$b_i = \text{Prob}\{\text{machine } i \text{ is blocked,}\} \qquad (2.7)$$
$$s_i = \text{Prob}\{\text{machine } i \text{ is starved,}\}$$

where $\delta \ll 1$;

(b) *each buffer is on the average close to being half full in the following sense:*

$$\mathrm{E}[h_i] = \frac{N_i}{2} \frac{N_i + 1}{N_i + 1 - p_i^f} + \mathcal{O}(\delta) \approx \frac{N_i}{2}, \quad i = 1, ..., M-1, \qquad (2.8)$$

where h_i is the steady state occupancy of the i-th buffer and $\mathrm{E}[\cdot]$ denotes the expectation.

Remark 2.1: Although condition (2.8) seems somewhat unexpected, it is, in retrospect, quite logical. Indeed, the buffer between machines i and $i+1$ is used to prevent the blockage of machine i and the starvation of machine $i+1$. Therefore, if this buffer is half full, it offers equal possibilities for alleviating the perturbations for both machines. If the buffer is too full, the production rate of machine i is too high, and p_i can be decreased and reallocated to another machine, possibly machine $i+1$, so that the production rate of the whole system is increased. Analogously, if buffer i is too empty, machine $i+1$ works "too fast" and a fraction of p_{i+1} can be transferred to another machine so that the production rate is increased. Thus, the status of the buffers – which ones are empty and which ones are full – offers guidance for potential improvements. Therefore, (2.8) is an **indicator of improvability with respect to WF**.

To characterize p_i^*, $i = 1, ..., M$, which render the system unimprovable, introduce

$$PR_{est}^* = \max_{\substack{p_1, ..., p_M \\ \prod_{i=1}^{M} p_i = p^*}} PR_{est}(p_1, ..., p_M, N_1, ..., N_{M-1}). \qquad (2.9)$$

Consider the following recursive procedure:

$$x(n+1) = (p^*)^{\frac{1}{M}} \prod_{i=1}^{M-1} \left(\frac{N_i + x(n)}{N_i + 1} \right)^{\frac{2}{M}}. \qquad (2.10)$$

Theorem 2.3 *Assume $\sum_{i=1}^{M-1} N_i^{-1} \leq M/2$. Then recursive procedure (2.10) is a contraction on $[0,1]$. Its steady state, x^*, satisfies the property*

$$x^* = \lim_{n \to \infty} x(n) = PR_{est}^*. \qquad (2.11)$$

Moreover, the sequence $p_1^*, ..., p_M^*$ which renders the serial production line (i)-(vi) unimprovable with respect to WF is defined by

$$\begin{aligned}
p_1^* &= \left(\frac{N_1 + 1}{N_1 + PR_{est}^*} \right) PR_{est}^*, \\
p_i^* &= \left(\frac{N_{i-1} + 1}{N_{i-1} + PR_{est}^*} \right) \left(\frac{N_i + 1}{N_i + PR_{est}^*} \right) PR_{est}^*, \quad i = 2, ..., M-1, \quad (2.12) \\
p_M^* &= \left(\frac{N_{M-1} + 1}{N_{M-1} + PR_{est}^*} \right) PR_{est}^*.
\end{aligned}$$

2.2.4. Improvability with respect to WF & WIP simultaneously

As it follows from Definition 2.3, serial production line (i)-(vi) is improvable with respect to WF & WIP simultaneously if there exist $p_1^*, ..., p_M^*$ and $N_1^*, ..., N_{M-1}^*$ such that $\prod_{i=1}^{M} p_i^* = p^*$ and $\sum_{i=1}^{M-1} N_i^* = N^*$ and

$$PR_{est}(p_1^*, ..., p_M^*, N_1^*, ..., N_{M-1}^*) > PR_{est}(p_1, ..., p_M, N_1, ..., N_{M-1}).$$

Theorem 2.4 *Let N^* be an integer multiple of $M-1$. Then serial production line (i)-(vi) is unimprovable with respect to WF & WIP simultaneously if and only if (2.6) takes place and, in addition,*

$$p_i^f = p_i^b, \quad i = 2, ..., M-1. \tag{2.13}$$

Corollary 2.2 *If condition (2.13) is satisfied, each intermediate machine is blocked and starved with practically the same frequency, i.e.,*

$$|b_i - s_i| \sim \mathcal{O}(\delta) \; i = 2, ..., M-1. \tag{2.14}$$

where b_i and s_i are defined in (2.7) and $\delta \ll 1$.

Remark 2.2: Condition (2.14) also can be given a simple interpretation. Buffers $i-1$ and i serve to prevent starvations and blockages of machine i, respectively. Thus, if machine i is blocked more often than it is starved, buffer $i-1$ can be reduced and the excess capacity could be added to another buffer, possibly buffer i. Analogously, if machine i is starved more often than blocked, N_i should be reduced and N_{i-1} increased. Conditions (2.8) and (2.14) are **indicators of improvability with respect to WF & WIP simultaneously.**

The values of N_i^* and p_i^* which render the system unimprovable with respect to WF & WIP simultaneously can be characterized as follows:

Theorem 2.5 *Let N^* be an integer multiple of $M-1$, and let*

$$PR_{est}^{**} = \max_{\substack{p_1, ..., p_M; \prod_{i=1}^{M} p_i = p^* \\ N_1, ..., N_{M-1}; \\ \sum_{i=1}^{M-1} N_i = N^*}} PR_{est}(p_1, ..., p_M, N_1, ..., N_{M-1}).$$

Then conditions (2.6) and (2.13) are satisfied if and only if

$$p_1^* = p_M^* = \left(\frac{N_1^* + 1}{N_1^* + PR_{est}^{**}}\right) PR_{est}^{**},$$

$$p_i^* = \left(\frac{N_i^* + 1}{N_i^* + PR_{est}^{**}}\right)^2 PR_{est}^{**}, \quad i = 2, ..., M - 1, \quad (2.15)$$

$$N_i^* = \frac{N^*}{M-1}, \quad i = 1, ..., M - 1.$$

Thus, a system is unimprovable with respect to WF & WIP simultaneously if and only if all buffers are of equal capacity. The isolation production rates of machines i, $i = 2, ..., M - 1$, are also the same; the isolation production rates of machines 1 and M are somewhat smaller than those for the other machines, as defined by the first two expressions in (2.15).

2.2.5. Improvability with respect to WIP

From Definition 2.1, serial production line (i)-(vi) is improvable with respect to WIP if there exists $N_1^*, ..., N_{M-1}^*$ such that $\sum_{i=1}^{M-1} N_i^* = N^*$ and

$$PR_{est}(p_1, ..., p_M, N_1^*, ..., N_{M-1}^*) > PR_{est}(p_1, ..., p_M, N_1, ..., N_{M-1}).$$

Theorem 2.6 *Serial production line (i)-(vi) is unimprovable with respect to WIP if and only if the quantity*

$$\min_{i=1,...,M} p_i \left(\min\left\{\frac{p_i^f}{p_i^b}, \frac{p_i^b}{p_i^f}\right\}\right) \quad (2.16)$$

is maximized over all sequences $N_1, ..., N_{M-1}$ such that $\sum_{i=1}^{M-1} N_i = N^*$.

Unfortunately, this result is of little practical significance, mainly due to the fact that the first and the last machines (which are never starved or blocked, respectively) are included in (2.16). In addition, the interpretation of (2.16) is much less obvious than that of (2.7). To alleviate these difficulties, we modify (2.16) to a form similar to (2.14) with the provision, however, that the p_i's, $i = 2, ..., M-1$, may not necessarily be equal to each other:

Criterion 2.1: Serial production line (i)-(vi) is practically unimprovable with respect to WIP if the quantity

$$\max_{i=2,\ldots,M-1} \frac{1}{p_i}|b_i - s_i| \qquad (2.17)$$

is minimized over all sequences N_i, $i = 1, \ldots, M-1$, such that $\sum_{i=1}^{M-1} N_i = N^*$, where b_i and s_i are defined in (2.7).

The term "practically unimprovable" is used here in the following sense:

Numerical Fact 2.1: Let $N_1^*, \ldots, N_{M-1}^*$ be the sequence which maximizes (2.16). Let $N_1^{**}, \ldots, N_{M-1}^{**}$ be the sequence which minimizes (2.17). Then

$$|PR_{est}(p_1, \ldots, p_M, N_1^*, \ldots, N_{M-1}^*) - PR_{est}(p_1, \ldots, p_M, N_1^{**}, \ldots, N_{M-1}^{**})| \ll 1.$$

This fact has been established on the basis of extensive computer simulation.

Remark 2.3: Criterion 2.1 can be given a simple interpretation: A system is practically unimprovable with respect to WIP if the weighted (by the inverse of the isolation production rates, $\frac{1}{p_i}$) differences between blockages and starvations for each intermediate machine are as close to each other as possible. This, in particular, implies that machines with smaller p_i's should have $|b_i - s_i|$ smaller than those with larger p_i's (protection of the "bottlenecks"). Condition (2.17), rewritten in the form

$$t_i = \frac{1}{p_i}|\tilde{b}_i - \tilde{s}_i| \approx \text{const}, \quad i = 2, \ldots, M-1, \qquad (2.18)$$

where $\tilde{b}_i = 1 - \frac{p_i^b}{p_i}$ and $\tilde{s}_i = 1 - \frac{p_i^f}{p_i}$, is referred to as an **indicator of improvability with respect to WIP**.

2.3. Constraint Relaxation

2.3.1. Bottleneck machine

Theorem 2.7 *In serial production lines unimprovable with respect to WF, the following property holds:*

$$p_i \frac{\partial PR_{est}}{\partial p_i} = \text{const}, \quad i = 1, \ldots, M. \qquad (2.19)$$

Relationship (2.19) implies that the machine with the smallest p_i corresponds to the largest $\frac{\partial PR_{est}}{\partial p_i}$. Therefore, in lines unimprovable with respect to WF the bottleneck machine can be found easily – it is the machine with the smallest production rate in isolation.

2.3.2. Bottleneck buffer

As defined in Section 2.1, buffer i^* is the bottleneck buffer of the line (i)-(vi) if
$$PR_{est}(p_1, ..., p_M, N_1, ..., N_{i^*}+1, ..., N_{M-1})$$
$$> PR_{est}(p_1, ..., p_M, N_1, ..., N_j+1, ..., N_{M-1}), \ \forall j \neq i^*.$$
As it is the case with the bottleneck machines, the smallest buffer is not necessarily the bottleneck. However, in lines unimprovable with respect to WIP the search for the bottleneck buffer becomes quite simple. Indeed, assume that a line is unimprovable in the sense that Procedure 3.1 of [4] has been carried out. Let machine i^* be the machine with the largest t_i. Then either buffer i^*-1 (if $p_{i^*}^b > p_{i^*}^f$) or buffer i^* (if $p_{i^*}^b < p_{i^*}^f$) is the bottleneck buffer in the following sense:

Numerical Fact 2.2: Let $N_1^*, ..., N_{M-1}^*$ be the assignment of N^* buffer spaces according to Procedure 3.1 of [4] and let machine i^* be the machine with the largest t_i. Let $N_1^{**}, ..., N_{M-1}^{**}$ be the assignment of N^*+1 buffer spaces according to Procedure 3.1 of [4]. Then
$$|PR_{est}(p_1, ..., p_M, N_1^*, ..., N_{i^*-1}^*+1, ..., N_{M-1}^*)$$
$$- PR_{est}(p_1, ..., p_M, N_1^{**}, ..., N_{M-1}^{**})| \ll 1,$$
if $p_{i^*}^b > p_{i^*}^f$, or
$$|PR_{est}(p_1, ..., p_M, N_1^*, ..., N_{i^*}^*+1, ..., N_{M-1}^*)$$
$$- PR_{est}(p_1, ..., p_M, N_1^{**}, ..., N_{M-1}^{**})| \ll 1,$$
if $p_{i^*}^b < p_{i^*}^f$.

This fact has been arrived at through numerical studies.

2.3.3. The choice of N^*

As it is clear from the above, an increase in N^* leads to an increase in PR_{est}. However, there exists an N^* such that any further increase in its value leads to an insignificant increase in PR_{est}. This phenomenon is captured in Definition 2.6 of N^* being ϵ-adapted to $p_1, ..., p_M$. For a serial production line (i)-(vi), N^* is ϵ-adapted to $p_1, ..., p_M$ if

$$\frac{|PR_{est}(N^*) - PR_{est}(N)|}{PR_{est}(N^*)} < \epsilon, \forall N \geq N^*,$$

where

$$PR_{est}(N^*) = PR_{est}(p_1, ..., p_M, N_1, ..., N_{M-1}),$$

$N^* = \sum_{i=1}^{M-1} N_i$, and $N_1, ..., N_{M-1}$ are calculated according to Procedure 3.1 of [4]. To characterize this N^*, define τ as the time that each part spends, on the average, from the moment it enters the first machine until the moment it is processed by the last machine. From Little's Law and Corollary 2.1, for a line (i)-(vi) unimprovable with respect to WF,

$$\tau \approx \frac{N^*}{2 PR_{est}(p_1^*, ..., p_M^*, N_1, ..., N_{M-1})}.$$

For a line (i)-(vi) unimprovable with respect to WIP,

$$\tau \leq \frac{N^*}{PR_{est}(p_1, ..., p_M, N_1^*, ..., N_{M-1}^*)} = T(N^*).$$

Since $PR_{est}(N^*)$ exhibits saturation, the increase of N^* beyond a certain value leads only to a practically linear increase of the residence time estimate $T(N^*)$ without a meaningful increase in PR_{est}. When the work-in-process is allowed to become infinite, the average production rate of the line becomes equal to the average production rate of the slowest machine, i.e. the machine with the smallest p_i. Thus, N^* which is ϵ-adapted to $p_1, ..., p_M$, can be determined from the equation

$$PR_{est}(N^*) = \frac{1}{1+\epsilon} \min_{i=1,...,M} p_i. \tag{2.20}$$

This N^*, being distributed according to Procedure 3.1 of [4], represents a compromise between the competing goals of ensuring a high average production rate and maintaining low work-in-process inventory.

3. Application

3.1. Machining Line

The theory described above has been applied to machining line of an automotive component plant. The structure of the line is shown in Fig. 1. Here each machine can be in either of the two states, up or down. In the up state, i-th machine is capable of producing at the rate C_i [parts/min]; when down, no production takes place. Machines are down due to cutting instrument changes and/or machine's breakdowns. The up- and down-time are random variables which, according to the available data, are distributed exponentially. Thus, the average up- and down-time of machine i are $\frac{1}{P_i}$ [min] and $\frac{1}{R_i}$ [min], $i = 1, ..., 8$, respectively, where P_i and R_i are the parameters of the exponential distributions.

Each buffer B_j, $j = 1, ..., 7$, is structured as a circular conveyer shown in Fig. 2; parts processed by machine i are placed on pallets circulating on a conveyer and are transported to machine $i+1$; then empty pallets are returned back to machine i.

Obviously, neither the machines nor the buffers satisfy the assumptions (i)–(vi). Therefore, to apply the improvability theory, these parameters have been massaged to result in p_i and N_i. These transformations are described below.

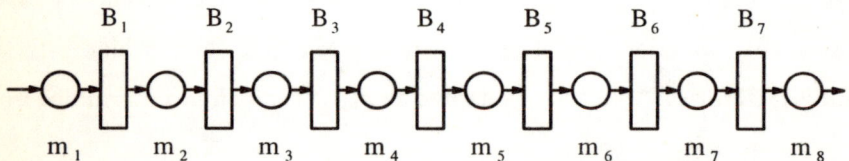

Figure 1: Serial production line.

3.2. Machines

In order to reduce the Markovian machine characterized by a triple (C_i, P_i, R_i) to a Bernoulli machine characterized by p_i, consider two serial production lines shown in Fig. 3. Here N_i^{Mar} and N_i^{Ber} are the capacity of the buffers in the Markovian and the Bernoulli lines, respectively. The following expressions have been derived heuristically to serve as a bridge between the Markovian and the Bernoulli models.

$$C_{max} = \max_{i=1,...,M}(C_i),$$

$$p_i = \frac{C_i \frac{R_i}{R_i+P_i}}{C_{max}}, \quad i = 1, ..., M,$$

$$N_i^{Ber} = \lceil \min(\frac{N_i^{Mar}}{C_{i+1}}R_i, \frac{N_i^{Mar}}{C_i}R_{i+1})\rceil, \quad i = 1, ..., M-1, \qquad (3.1)$$

where $\lceil z \rceil$ is the smallest integer greater than z.

The accuracy of this transformation has been tested numerically. For Markovian lines which result in $N_i^{Ber} \geq 2$, the error in the production rate estimation using the two models is small (less than 1%). For $N_i^{Ber} = 1$, the error is substantial (5-10%). However, given the imprecision of the initial data available on the factory floor (typically, 5-10% as well), we nevertheless utilize (3.1) as a tool for connecting the Markovian and the Bernoulli models of serial production lines.

3.3. Buffers

The structure of the buffer between each of the eight operations is shown in Fig. 2. The conveyers operate as follows: When an empty pallet reaches machine i, it is stopped and held there until a part is placed. When the full pallet reaches machine $i+1$, it is again stopped until the part is picked up (by a robotic device) and placed for processing at machine $i+1$. Thus, the buffer capacity of the circular conveyer between machine i and $i+1$ can be expressed as

$$N_i^{Mar} = \{\text{Total number of pallets on the circular conveyer}$$
$$- \text{ Number of pallets required to sustain the}$$
$$\text{production rate of } \min(C_i, C_{i+1})\}, \ i = 1, ..., \ 7,$$

where the number of pallets required to sustain the production rate of $\min(C_i, C_{i+1})$ is calculated based on the speed and length of conveyer and the values of C_i and C_{i+1}.

3.4. Parameters Identified

Based upon the approach outlined above, the parameters of the eight operations and 7 buffers involved have been identified. They are given in Table 1 for the Markovian representation and in Table 2 for the equivalent Bernoulli representation. These parameters are used below for the performance and improvability analysis.

Table 1: Markovian machine parameters identified

i	1	2	3	4	5	6	7	8
P_i	0.0041	0.0041	0.0935	0.0714	0.1704	0.0025	0.1235	0.0532
R_i	0.0099	0.0099	0.2882	03774	0.3030	0.0333	0.3846	0.4167
C_i	21.6	11.33	12.25	10.25	14.1	19	12.12	9.56
N_i^{Mar}	10360	100	40	50	100	50	50	—

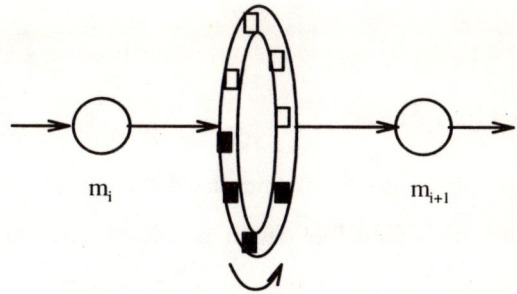

■ : Pallet with a part   □ : Empty pallet

Figure 2: Circular conveyer.

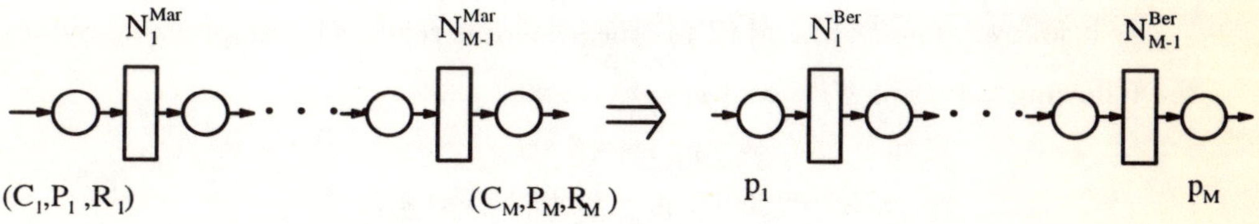

Figure 3: Markovian and corresponding Bernoulli lines.

Table 2: Bernoulli machine parameters calculated from Table 1

i	1	2	3	4	5	6	7	8
p_i	0.71	0.37	0.43	0.40	0.42	0.82	0.43	0.39
N_i^{Ber}	5	1	2	2	1	1	2	—

3.5. Production Rate

Using the recursive procedure (2.3) and the data given in Table 2, the production rate of the machining line under consideration has been calculated to be 4.71 [parts/min]. Using the aggregation procedure for Markovian lines described in [4], and the data of Table 1, the production rate has been calculated to be 4.55 [parts/min]. The error, 3%, indicates an acceptable accuracy of the production rate estimation using the Bernoulli and the Markovian descriptions. What seems to be unacceptable is the system production rate, 4.55 [parts/min], which is about one half of the production rate of the worst operation (machine 2), which could produce at the rate $C_2 \frac{R_2}{P_2+R_2} = 8.04$ [parts/min].

3.6. Improvability Under Constraints

Using the data of Table 2, introduce

$$N_\Sigma = \sum_{i=1}^{7} N_i^{Ber} = 14,$$
$$p_\Pi = \prod_{i=1}^{8} p_i = 0.0026. \quad (3.2)$$

As it follows from (2.8) and (2.14) the system is rendered unimprovable by the following values of N_i^{Ber} and p_i:

$$N_i^{Ber} = 2, \quad i = 1, ..., 7,$$
$$p_1 = p_8 = 0.3893, \quad p_i = 0.5082, \quad i = 2, ..., 7. \quad (3.3)$$

With these parameters, the behavior of the system is characterized by $PR = 6.44$ [parts/min], which is 37% improvement as compared with the initial performance. The implementation of the improvement measures (3.3), however, is impossible due to the system structure: the capacity of N_1^{Ber}

physically cannot be reallocated to any of $N_i^{Ber}, i = 2, ..., 7$. Therefore, in practical terms, improvability of system shown in Fig. 1 can be accomplished only by a relaxation of constraints (3.2), i.e., an increase of N_Σ and p_Π and an appropriate allocation of the additional WIP and WF.

3.7. Constraints Relaxation: Increase of N_Σ

Technically, an increase of N_Σ can be accomplished by placing a loading-unloading robot next to each of the circular conveyers. When machine $i+1$ is down, the robot unloads the parts from the conveyer thus preventing the blockage of machine i. When machine i is down, the robot places the parts on the conveyer thus preventing the starvation of machine $i+1$. The capacity of such a robotic material handling system is quite large and for all practical purposes equals infinity. Since the robots required for such an automation are very expensive, the problem is to find the smallest number and the location of the robots that optimize the production rate.

To solve this problem, an improvability indicator (2.16) has been used. The results are shown in Tables 3 along with the production rate corresponding to each type of the robot allocation. As it follows from these data, the number of robots that lead to the practically optimal production rate is 5, and their locations are at $B_2, B_3, B_4, B_6,$ and B_7. With this robots, the system production rate is improved to 7.93 [parts/min], i.e., by 68%.

3.8. Constraint Relaxation: Increase in p_Π

A further improvement of the system operation can be achieved by increasing p_Π. Table 4 gives the production rate with p_Π increased by 10% and 20% and allocated according to the improvability indicator (2.8). As it follows from this table, with both, 5 robots installed and p_Π increased by 10% (by improved preventive maintenance procedures), the production rate of the line can be improved by 95%.

4. Conclusions

This paper describes the production system research that has been carried out at Michigan for the last 10 years. We would like to see this as a contribution to the engineering knowledge concerning the production systems. The results derived so far can be viewed somewhat akin to the laws of statics in mechanics. Formulated in general terms they read:

(α) In a well designed production system each buffer is, on the average, half full.

(β) In a well designed production system all intermediate machines have equal probability of blockage and starvation.

(γ) In a well designed system the maximal WIP is adapted to the machines' isolation production rates.

Generalization of these laws for other types of production systems as well as the investigation of the laws of dynamics are the topics of the future work.

REFERENCES

1. E. Deming, *Out of the Crisis*, Massachusetts Institute of Technology, Cambridge, Massachusetts, (1986).

2. E. Goldratt and J. Cox, *The Goal*, North Rivers Press, New York, (1986).

3. G. Taguchi, *System of Experimental Design: Engineering Methods to Optimize Quality and Minimize Cost*, UNIPUB/Kraus International Publications, Dearborn, Michigan (1987).

4. D. Jacobs and S.M. Meerkov, "Mathematical Theory of Improvability for Production Systems", *Mathematical Problems in Engineering*, **1**(2), 95-137 (1995).

5. J. A. Buzacott and L. E. Hanifin, "Models of Automatic Transfer Lines With Inventory Banks - A Review and Comparison," *AIIE Trans.*, **10**, 197-207 (1978).

6. J.A. Buzacott and J.G. Shanthikumar, *Stochastic Models of Manufacturing Systems*, Prentice Hall, (1993).

7. Y. Dallery and S.B. Gershwin, Manufacturing Flow Line Systems: A Review of Models and Analytical Results, *Queu. Syst.*, **12**, 3-94 (1992).

8. S.B. Gershwin, *Manufacturing Systems Engineering*, Prentice Hall, (1994).

D. JACOBS
FORD MOTOR COMPANY, EFHD
MCKEAN AND TEXTILE ROADS, P.O.BOX 922
YPSILANTI, MI 48197

C.-T. KUO
DIVISION OF SYSTEMS SCIENCE
DEPARTMENT OF ELECTRICAL ENGINEERING AND
COMPUTER SCIENCE
THE UNIVERSITY OF MICHIGAN
ANN ARBOR, MI 48109-2122

J.-T. LIM
DIVISION OF SYSTEMS SCIENCE
DEPARTMENT OF ELECTRICAL ENGINEERING AND
COMPUTER SCIENCE
THE UNIVERSITY OF MICHIGAN
ANN ARBOR, MI 48109-2122
 ON LEAVE FROM
DEPARTMENT OF ELECTRICAL ENGINEERING
KOREA ADVANCED INSTITUTE OF SCIENCE AND
TECHNOLOGY
TAEJON, KOREA

S.M. MEERKOV
DIVISION OF SYSTEMS SCIENCE
DEPARTMENT OF ELECTRICAL ENGINEERING AND
COMPUTER SCIENCE
THE UNIVERSITY OF MICHIGAN
ANN ARBOR, MI 48109-2122

Table 3: Production rates corresponding to each type of the robot allocation

No. of robots installed	Optimal allocation of installing robots	Production rate [parts/min]	% of improvement
1	B_2	5.29	12%
2	B_2, B_6	6.00	27%
3	B_2, B_4, B_6	6.76	44%
4	B_2, B_4, B_6, B_7	6.91	47%
5	B_2, B_3, B_4, B_6, B_7	7.93	68%
6	$B_2, B_3, B_4, B_5, B_6, B_7$	7.99	70%
7	$B_1, B_2, B_3, B_4, B_5, B_6, B_7$	7.99	70%

Table 4: Production rates of the system with p_Π increase

p_Π improvement	Production rate [parts/min]	% of improvement
10 %	9.18	95 %
20 %	9.29	97 %

Part III

Approximate Optimality and Robustness

Asymptotic Optimal Feedback Controls in Stochastic Dynamic Two-Machine Flowshops*

Suresh P. Sethi and Xun Yu Zhou

Abstract

In this paper, we treat an optimal control problem of a stochastic two-machine flowshop with machines subject to random breakdown and repair. While the problem is difficult to solve, it can be approximated by a deterministic problem when the rates of machine failure and repair become large. Furthermore, beginning with the solution of the deterministic problem, we can construct a feedback control for the stochastic flowshop that is asymptotically optimal with respect to the rates of changes in machine states. We also show that a threshold type control known also as Kanban control is asymptotically optimal in some cases and not in others.

1 Introduction

We are concerned with the problem of controlling the production rate of a stochastic two-machine flowshop in order to meet a constant demand facing the flowshop at a minimum expected discounted cost of inventory/shortage. The stochastic nature of the flowshop is due to the machines that are subject to random breakdown and repair. In addition to the states of machines, we take the inventory of the unfinished goods, known as work-in-process, and the difference of the cumulative production of finished goods and the cumulative demand, known as surplus, as state variables. It is important to point out that the work-in-process in the internal buffer must remain nonnegative. Also, there may be a limitation on the size of the internal buffer.

---

*This research was supported in part by NSERC Grant A4169, MRCO, CUHK Direct Grant 93-94, RGC Earmarked Grant CUHK 249/94E, and RGC Earmarked Grant CUHK 489/95E.

This problem is one of the simplest optimization problems concerning stochastic manufacturing systems that involve an internal buffer linking a pair of machines. Yet the problem is not easy to solve because of the state constraint associated with the buffer. Certainly no explicit solution, unlike in the single machine system without any state constraint [2], is available for the stochastic two-machine flowshop.

However, the problem is extremely important because it captures the essence of many realistic manufacturing systems, especially those producing semiconductors [20, 21]. These systems have machines that are connected together by internal buffers and are subject to associated state constraints. Furthermore, semiconductor manufacturing involves failure-prone machines modelled here as finite state Markov processes. Therefore, substantial efforts have been made to solve the problem computationally to obtain optimal or near-optimal controls [1, 3, 6, 8, 11]. Theoretical research has also been conducted to study the dynamic properties of the system and to obtain appropriate boundary conditions for the Bellman equation of the problem [9, 10]. Lou, Sethi and Zhang [9] also provide a partial characterization of optimal controls.

Recognizing the complexity of the problem, Sethi, Zhang and Zhou [15] have developed a hierarchical approach to solving the stochastic optimal control problem of an N-machine flowshop approximately, when the rates of changes in machine states are much larger than the rate of discounting costs. They show that the stochastic problem can be approximated by a deterministic optimization problem. Then by using the optimal control of the deterministic problem, they construct an open-loop control for the original stochastic problem, which is asymptotically optimal as the rates of changes in machine states approach infinity.

While asymptotically optimal, the constructed open-loop controls are not expected to perform well unless the rates of changes in machine states are unrealistically large. What is required therefore is a construction of asymptotic optimal *feedback* controls. At present, such constructions are available only for single or parallel machine systems [14, 16], in which no state constraints are present. In such systems, either the Lipschitz property of optimal control for the corresponding deterministic systems or the monotonicity property of optimal control with respect to the state variables makes the proof of asymptotic optimality go through. Unfortunately, these properties are no longer available in the case of flowshops.

It is the purpose of this paper to construct asymptotic optimal feedback controls for the two-machine flowshop under consideration. Based on the explicit characterization of optimal controls for the deterministic problem [18], we are able to construct approximately optimal feedback controls for the stochastic problem and prove their asymptotic optimality. It should be noted that Samaratunga, Sethi and Zhou [12] have already made a computational investigation of such constructed controls and found them to perform very well in comparison to other existing methods in the literature [11, 13].

The plan of this paper is as follows. In the next section we formulate the stochastic problem of the two-machine flowshop, obtain the corresponding deterministic problem, and review relevant results that are needed in our analysis. In Section 3 we construct feedback controls and prove their asymptotic optimality in the typical case when the average maximum capacity of the first machine is not less than that of the second. For a heuristic explanation of their asymptotic optimality, see [12]. Section 4 shows that the so-called Kanban controls adapted to the stochastic flowshop are asymptotically optimal in some cases and *not* in others. Section 5 concludes the paper along with some suggestions for future research.

2 The Model and a Review of Relevant Results

We consider a dynamic stochastic flowshop consisting of two unreliable machines M_1 and M_2 shown in Fig. 1. The machines are subject to breakdown and repair. Each machine is assumed to have two states: up and down. We

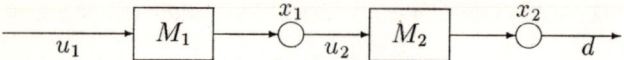

Figure 1: A two-machine flowshop

assume that the first and the second machines have maximum production capacities m_1 and m_2, respectively. Therefore, the system has four machines states: $\mathbf{k}^1 = (m_1, m_2)$ corresponds to both machines up, $\mathbf{k}^2 = (m_1, 0)$ to M_1 up and M_2 down, $\mathbf{k}^3 = (0, m_2)$ to M_1 down and M_2 up, and $\mathbf{k}^4 = (0,0)$ to both machines down. Let $\mathcal{M} = \{\mathbf{k}^1, \mathbf{k}^2, \mathbf{k}^3, \mathbf{k}^4\}$. We denote the production capacity process by $\mathbf{k}(\varepsilon, t) = (k_1(\varepsilon, t), k_2(\varepsilon, t))$, where ε is a small parameter to be specified later. Here and elsewhere we use boldface letters to stand for vectors (e.g., $\mathbf{x} = (x_1, x_2)$, $\mathbf{u} = (u_1, u_2)$, $\boldsymbol{\nu} = (\nu_1, \nu_2, \nu_3, \nu_4)$, etc.).

We use $u_1(t)$ and $u_2(t)$, $t \geq 0$, to denote the production rates on the first and the second machine, respectively. We denote the work-in-process between M_1 and M_2 as $x_1(t)$ (with $0 \leq x_1(t) \leq B$) and the surplus level of the finished goods as $x_2(t)$. Here B represents the size of the internal buffer. On the other hand, a positive surplus means inventory and a negative surplus means shortage. With an assumed constant demand rate d for finished goods, the dynamics of the system can be written as follows:

$$\begin{cases} \dot{x}_1(t) = u_1(t) - u_2(t), & x_1(0) = x_1 \\ \dot{x}_2(t) = u_2(t) - d, & x_2(0) = x_2 \end{cases}, \quad (2.1)$$

where

$$0 \leq u_i(t) \leq k_i(\varepsilon, t), \ t \geq 0, \ i = 1, 2. \quad (2.2)$$

Notation

We make use of the following additional notation in this paper:

b^+ : $= \max\{b, 0\}$ for any real number b;

b^- : $= \max\{-b, 0\}$ for any real number b;

$|(b_1, \cdots, b_n)|$: $= |b_1| + \cdots + |b_n|$ for any vector $(b_1, \cdots, b_n)$ with any positive integer n;

χ_A : the indicator function of any set A;

$P(A_1; A_2; \cdots, A_m)$: $= \text{Prob}(A_1 \cap A_2 \cap \cdots \cap A_m)$;

$E(\xi; A)$: $= E(\xi \cdot \chi_A)$ for any random variable ξ and any set A;

$O(y)$: a variable such that $\sup_y |O(y)/y| < \infty$;

$C^1(A)$: set of continuously differentiable functions defined on a set A.

Let $S = [0, B] \times R^1 \subset R^2$ denote the state constraint domain. We can now define the set of admissible controls $\mathbf{u}(\cdot) = (u_1(\cdot), u_2(\cdot))$.

Definition 2.1

We say that a control $\mathbf{u}(\cdot) = (u_1(\cdot), u_2(\cdot))$ is *admissible* with respect to the initial state value $\mathbf{x} = (x_1, x_2) \in S$ if (i) $\mathbf{u}(t)$ is adapted to $\mathcal{F}_t^\varepsilon = \sigma\{\mathbf{k}(\varepsilon, s) : 0 \leq s \leq t\}$, the sigma algebra generated by the process $\mathbf{k}(\varepsilon, s)$, $0 \leq s \leq t$, (ii) $0 \leq u_i(t) \leq k_i(\varepsilon, t)$ for $t \geq 0$ and $i = 1, 2$, and (iii) the corresponding state

$$\mathbf{x}(t) = (x_1(t), x_2(t)) \in S \text{ for all } t \geq 0. \tag{2.3}$$

Definition 2.2

We say that a function $\mathbf{u} = S \times \mathcal{M} \to R^2$ is an *admissible feedback control* if (i) for any given initial $\mathbf{x} = (x_1, x_2) \in S$, the following equation has a unique solution $\mathbf{x}(\cdot)$:

$$\begin{cases} \dot{x}_1(t) = u_1(\mathbf{x}(t), \mathbf{k}(\varepsilon, t)) - u_2(\mathbf{x}(t), \mathbf{k}(\varepsilon, t)), & x_1(0) = x_1 \\ \dot{x}_2(t) = u_2(\mathbf{x}(t), \mathbf{k}(\varepsilon, t)) - d, & x_2(0) = x_2 \end{cases},$$

and (ii) $\mathbf{u}(\mathbf{x}(\cdot), \mathbf{k}(\varepsilon, \cdot))$ is admissible with respect to $\mathbf{x}$.

Our problem is to find an admissible control $\mathbf{u}(\cdot)$ that minimizes the cost function

$$J^\varepsilon(\mathbf{x}, \mathbf{k}, \mathbf{u}(\cdot)) = E \int_0^\infty e^{-\rho t}(c_1 x_1(t) + c_2^+ x_2^+(t) + c_2^- x_2^-(t))dt, \qquad (2.4)$$

where c_1, c_2^+ and c_2^- are given *nonnegative* constants and $\mathbf{k} = (k_1, k_2)$ is the initial value of $\mathbf{k}(\varepsilon, t)$. For a feedback control $\mathbf{u}$, on the other hand, we shall write the cost $J^\varepsilon(\mathbf{x}, \mathbf{k}, \mathbf{u}(\mathbf{x}(\cdot), \mathbf{k}(\varepsilon, \cdot)))$ as $J^\varepsilon(\mathbf{x}, \mathbf{k}, \mathbf{u})$, where $\mathbf{x}(\cdot)$ is the corresponding trajectory under $\mathbf{u}$ with the initial state $\mathbf{x}$ and the initial machine state $\mathbf{k}$.

We use $\mathcal{A}^\varepsilon(\mathbf{x})$ to denote the set of admissible controls with respect to the initial state $\mathbf{x} \in S$ and the initial machine state $\mathbf{k}$, and $v^\varepsilon(\mathbf{x}, \mathbf{k})$ to denote the value function, i.e.,

$$v^\varepsilon(\mathbf{x}, \mathbf{k}) = \inf_{\mathbf{u}(\cdot) \in \mathcal{A}^\varepsilon(\mathbf{x})} J^\varepsilon(\mathbf{x}, \mathbf{k}, \mathbf{u}(\cdot)). \qquad (2.5)$$

Assumptions

We make the following assumptions:

(A1) The capacity process $\mathbf{k}(\varepsilon, t) \in \mathcal{M}$ is a finite state Markov chain with generator $Q^\varepsilon = \varepsilon^{-1}Q$, where $Q = (q_{ij})$ is a 4×4 matrix with $q_{ij} \geq 0$ if $j \neq i$ and $q_{ii} = -\sum_{j \neq i} q_{ij}$. Moreover, Q is irreducible and is taken to be the one that satisfies

$$\min_{ij}\{|q_{ij}| : q_{ij} \neq 0\} = 1.$$

Let $\boldsymbol{\nu} = (\nu^1, \nu^2, \nu^3, \nu^4) > 0$, the only positive solution of

$$\boldsymbol{\nu} Q = 0 \text{ and } \sum_{j=1}^{4} \nu^j = 1,$$

denote the equilibrium distribution of Q.

(A2) $a_1 \geq a_2 \geq d$, where $\mathbf{a} = (a_1, a_2)$ denotes the average maximum capacities of M_1 and M_2 respectively, i.e., $a_1 = (\nu^1 + \nu^2)m_1$ and $a_2 = (\nu^1 + \nu^3)m_2$.

We use $\mathcal{P}^\varepsilon$ to denote the stochastic optimal control problem facing the flowshop, i.e.,

$$\mathcal{P}^\varepsilon : \begin{cases} \text{minimize} & J^\varepsilon(\mathbf{x}, \mathbf{k}, \mathbf{u}(\cdot)) = E \int_0^\infty e^{-\rho t} h(\mathbf{x}(t)) dt \\ \text{subject to} & \begin{cases} \dot{x}_1(t) = u_1(t) - u_2(t), \quad x_1(0) = x_1, \\ \dot{x}_2(t) = u_2(t) - d, \quad x_2(0) = x_2, \\ \mathbf{u}(\cdot) \in \mathcal{A}^\varepsilon(\mathbf{x}) \end{cases} \\ \text{value function} & v^\varepsilon(\mathbf{x}, \mathbf{k}) = \inf_{\mathbf{u}(\cdot) \in \mathcal{A}^\varepsilon(\mathbf{x})} J^\varepsilon(\mathbf{x}, \mathbf{k}, \mathbf{u}(\cdot)), \end{cases}$$

where we have used an abbreviated notation $h(\mathbf{x}) = c_1 x_1 + c_2^+ x_2^+ + c_2^- x_2^-$.

Intuitively, as the rates of breakdown and repair of the machines approach infinity, $\mathcal{P}^\varepsilon$, termed the *original problem*, can be approximated by a simpler deterministic problem called the *limiting problem*, in which the stochastic machine capacity process $\mathbf{k}(\varepsilon, t)$ is replaced by its average value. To define the limiting problem, we consider the following class of deterministic controls.

Definition 2.3

For $\mathbf{x} \in S$, let $\bar{\mathcal{A}}(\mathbf{x})$ denote the set of the deterministic measurable controls

$$\mathbf{U}(\cdot) = (\mathbf{u}^1(\cdot), \cdots, \mathbf{u}^4(\cdot)) = ((u_1^1(\cdot), u_2^1(\cdot)), \cdots, (u_1^4(\cdot), u_2^4(\cdot)))$$

such that $0 \leq u_i^j(t) \leq k_i^j$ for $t \geq 0$, $i = 1, 2$ and $j = 1, 2, 3, 4$, and that the corresponding solutions $\mathbf{x}(\cdot)$ of the following system

$$\begin{cases} \dot{x}_1(t) = \sum_{j=1}^4 \nu^j u_1^j(t) - \sum_{j=1}^4 \nu^j u_2^j(t), & x_1(0) = x_1, \\ \dot{x}_2(t) = \sum_{j=1}^4 \nu^j u_2^j(t) - d, & x_2(0) = x_2. \end{cases} \quad (2.6)$$

satisfy $\mathbf{x}(t) \in S$ for all $t \geq 0$.

We can now specify the limiting problem denoted as

$$\bar{\mathcal{P}}: \begin{cases} \text{minimize} & J(\mathbf{x}, \mathbf{U}(\cdot)) = \int_0^\infty e^{-\rho t} h(\mathbf{x}(t)) dt \\ \text{subject to} & \begin{cases} \dot{x}_1(t) = \sum_{j=1}^4 \nu^j u_1^j(t) - \sum_{j=1}^4 \nu^j u_2^j(t), & x_1(0) = x_1, \\ \dot{x}_2(t) = \sum_{j=1}^4 \nu^j u_2^j(t) - d, & x_2(0) = x_2, \\ \mathbf{U}(\cdot) \in \bar{\mathcal{A}}(\mathbf{x}) \end{cases} \\ \text{value function} & v(\mathbf{x}) = \inf_{\mathbf{U}(\cdot) \in \bar{\mathcal{A}}(\mathbf{x})} J(\mathbf{x}, \mathbf{U}(\cdot)). \end{cases}$$

Let

$$w_1(t) = \sum_{j=1}^4 \nu^j u_1^j(t) \text{ and } w_2(t) = \sum_{j=1}^4 \nu^j u_2^j(t). \tag{2.7}$$

Then the limiting problem $\bar{\mathcal{P}}$ can be transformed to

$$\bar{\mathcal{P}}: \begin{cases} \text{minimize} & J(\mathbf{x}, \mathbf{w}(\cdot)) = \int_0^\infty e^{-\rho t} h(\mathbf{x}(t)) dt \\ \text{subject to} & \begin{cases} \dot{x}_1(t) = w_1(t) - w_2(t), & x_1(0) = x_1, \\ \dot{x}_2(t) = w_2(t) - d, & x_2(0) = x_2, \\ \mathbf{w}(\cdot) \in \bar{\mathcal{A}}(\mathbf{x}) \end{cases} \\ \text{value function} & v(\mathbf{x}) = \inf_{\mathbf{w}(\cdot) \in \bar{\mathcal{A}}(\mathbf{x})} J(\mathbf{x}, \mathbf{w}(\cdot)), \end{cases}$$

where $\bar{\mathcal{A}}(\mathbf{x})$, $\mathbf{x} \in S$, can also be considered, with an abuse of notation, to be the set of the deterministic measurable controls $\mathbf{w}(\cdot) = (w_1(\cdot), w_2(\cdot))$ with

$$0 \leq w_1(t) \leq a_1, \quad 0 \leq w_2(t) \leq a_2,$$

and with the corresponding solution $\mathbf{x}(\cdot)$ of the state equation appearing in $\bar{\mathcal{P}}$ satisfying $\mathbf{x}(t) \in S$ for all $t \geq 0$.

The next result is standard and well known.

Theorem 2.1

The value functions v^ε and v are convex in $\mathbf{x}$.

The following theorem is a special case of the results proved in Sethi, Zhang and Zhou [17], and it says that the problem $\bar{\mathcal{P}}$ is indeed a limiting problem in the sense that the value function v^ε of $\mathcal{P}^\varepsilon$ converges to the value function v of $\bar{\mathcal{P}}$. Moreover, it gives the corresponding convergence rate.

Theorem 2.2

There exists a positive constant C such that

$$|v^\varepsilon(\mathbf{x},\mathbf{k}) - v(\mathbf{x})| \leq C(1+|\mathbf{x}|)\varepsilon^{\frac{1}{3}}, \qquad (2.8)$$

for $\mathbf{x} \in S$ and sufficiently small ε.

Next, we define what are asymptotic optimal feedback controls and give some estimates related to the behavior of the process $\mathbf{k}(\varepsilon, t)$ at small ε.

Definition 2.4

An admissible feedback control $\mathbf{u}^\varepsilon(\mathbf{x},\mathbf{k})$ is *asymptotic optimal* if

$$\lim_{\varepsilon \to 0} |J^\varepsilon(\mathbf{x},\mathbf{k},\mathbf{u}^\varepsilon) - v^\varepsilon(\mathbf{x},\mathbf{k})| = 0.$$

Moreover, if there exist positive constants $C(\mathbf{x})$ and γ such that

$$|J^\varepsilon(\mathbf{x},\mathbf{k},\mathbf{u}^\varepsilon) - v^\varepsilon(\mathbf{x},\mathbf{k})| \leq C(\mathbf{x})\varepsilon^\gamma,$$

then $C(\mathbf{x})\varepsilon^\gamma$ is called an *asymptotic error bound* and γ is called the *rate of convergence*.

Lemma 2.1

For any bounded deterministic measurable process $\beta(t)$, there exists a positive constant C such that

$$E|\int_0^t (\chi_{\{\mathbf{k}(\varepsilon,s)=\mathbf{k}^i\}} - \nu^i)\beta(s)ds|^2 \leq C\varepsilon(1+t^2), \ t \geq 0, \ i = 1,2,3,4,$$

for sufficiently small ε.

Proof

See Sethi, Zhang and Zhou [16].

Lemma 2.2

For any bounded deterministic measurable process $\beta(t)$, there exist positive constants C and K such that

$$P(|\int_0^t (\chi_{\{\mathbf{k}(\varepsilon,s)=\mathbf{k}^i\}} - \nu^i)\beta(s)ds| \geq \varepsilon^{\frac{1}{3}})$$
$$\leq C(e^{-K\varepsilon^{-1}(1+t)} + e^{-K\varepsilon^{-1/3}(1+t)^{-3}}), \ t \geq 0, \ i = 1,2,3,4,$$

for sufficiently small ε.

Proof

This is a special case of Lemma 4.3 in Sethi, Zhang and Zhou [15] with $\sigma = \frac{1}{3}$.

Corollary 2.1

Let τ be a bounded $\mathcal{F}_t^\varepsilon$–stopping time with $\tau \leq a$, P-a.s. Then

$$P(|\int_0^\tau (\chi_{\{\mathbf{k}(\varepsilon,s)=\mathbf{k}^i\}} - \nu^i)\beta(s)ds| \geq \varepsilon^{\frac{1}{3}})$$
$$\leq C(e^{-K\varepsilon^{-1}} + e^{-K\varepsilon^{-1/3}(1+a)^{-3}}), \ t \geq 0, \ i = 1,2,3,4.$$

Proof

Since τ is a *bounded* stopping time, all the arguments in proving Lemma 4.3 in Sethi, Zhang and Zhou [15] go through by virtue of the Optional Sampling Theorem of Martingales (see, for example, Ikeda and Watanabe [7, Theorem I.6.1].

The purpose of this paper is to construct asymptotic optimal controls for our original problem $\mathcal{P}^\varepsilon$. The construction will begin with the optimal controls for the limiting deterministic problem $\bar{\mathcal{P}}$. Explicit optimal feedback controls for $\bar{\mathcal{P}}$ have been obtained in Sethi and Zhou [18], and are recapitulated below for convenience in exposition.

Case I: $c_1 < c_2^+$

The optimal feedback control is given by

$$\mathbf{w}^*(\mathbf{x}) = \begin{cases} (d,d), & x_1 = x_2 = 0, \\ (0,0), & 0 \le x_1 \le B, x_2 > 0, \\ (0,d), & 0 < x_1 \le B, x_2 = 0, \\ (0,a_2), & 0 < x_1 \le B, x_2 < 0, \\ (a_2,a_2), & x_1 = 0, x_2 < 0. \end{cases} \qquad (2.9)$$

Case II: $c_1 \ge c_2^+$

The optimal feedback control is given by

$$\mathbf{w}^*(\mathbf{x}) = \begin{cases} (d,d), & x_1 = x_2 = 0, \\ (0,a_2), & 0 < x_1 \le B, \\ (0,0), & x_1 = 0, x_2 > 0, \\ (a_2,a_2), & x_1 = 0, x_2 < 0. \end{cases} \qquad (2.10)$$

3 Construction of Asymptotic Optimal Controls for the Stochastic Problem $\mathcal{P}^\varepsilon$

In this section, we construct a feedback control for $\mathcal{P}^\varepsilon$ along with a rigorous argument for its asymptotic optimality. The main idea is to use a control for the original stochastic problem $\mathcal{P}^\varepsilon$ that *has the same form* as an approximately constructed near-optimal feedback control for the limiting deterministic problem $\bar{\mathcal{P}}$, and then to show that the two trajectories of $\mathcal{P}^\varepsilon$ and $\bar{\mathcal{P}}$ under their respective controls are very close to each other *on average*.

3.1 The Case $c_1 < c_2^+$

The optimal control $\mathbf{w}^*$ for the limiting problem $\bar{\mathcal{P}}$ when $c_1 < c_2^+$ is presented in (2.9). Clearly, this control is not feasible for the stochastic problem $\mathcal{P}^\varepsilon$ when one or both of the machines are broken down. Since our purpose here is to construct an asymptotic optimal control for $\mathcal{P}^\varepsilon$ beginning with $\mathbf{w}^*$, we first set $\mathbf{w}^*(\mathbf{x}) = \mathbf{r}^*(\mathbf{x}, \mathbf{a})$, where

$$\mathbf{r}^*(\mathbf{x}, \mathbf{k}) = \begin{cases} (\min\{k_1, k_2, d\}, \min\{k_1, k_2, d\}), & x_1 = x_2 = 0, \\ (0, 0), & 0 \leq x_1 \leq B, x_2 > 0, \\ (0, \min\{k_2, d\}), & 0 < x_1 \leq B, x_2 = 0, \\ (0, k_2), & 0 < x_1 \leq B, x_2 < 0, \\ (\min\{k_1, k_2\}, \min\{k_1, k_2\}), & x_1 = 0, x_2 < 0. \end{cases} \quad (3.1)$$

We should note that the function $\mathbf{r}^*(\mathbf{x}, \mathbf{k})$ is constructed so that $\mathbf{w}^*(\mathbf{x}) = \mathbf{r}^*(\mathbf{x}, \mathbf{a})$ is optimal for $\bar{\mathcal{P}}$ when $a_1 \geq a_2$ and $c_1 < c_2^+$. In addition, if we use the control $\mathbf{u}(\mathbf{x}, \mathbf{k}) = \mathbf{r}^*(\mathbf{x}, \mathbf{k})$ for $\mathcal{P}^\varepsilon$, then in the interior of the state space S, it uses the maximum (resp. minimum) available machine capacity on M_i whenever the capacity a_i (resp. 0) is used in $\bar{\mathcal{P}}$ on the corresponding deterministic machine $M_i, i = 1, 2$. Furthermore, it tries in the interior of S to produce on M_i as much, but no more than d, as possible, whenever the production rate is d on the corresponding deterministic machine $M_i, i = 1, 2$. Finally, along the constraint boundary $x_1 = 0$, it tries to use the above two rules as much as possible and still remain feasible. Clearly, the construction of $\mathbf{r}^*(\mathbf{x}, \mathbf{k})$ is derived from the insight that in the absence of any production costs, the optimal control in $\bar{\mathcal{P}}$ is a combination of bang-bang and singular controls that is directed to achieve the buffer levels $x_1 = x_2 = 0$.

Despite all this effort, however, our task is not quite finished yet, as one can easily see that $\mathbf{u}^*(\mathbf{x}, \mathbf{k}) = \mathbf{r}^*(\mathbf{x}, \mathbf{k})$ is *not* asymptotic optimal for the original problem $\mathcal{P}^\varepsilon$, when the unit shortage cost c_2^- is strictly positive. This is because when the trajectory under this control reaches $(0, x_2), x_2 < 0$, it moves along the line $x_1 = 0$ at a different average rate than the rate in $\bar{\mathcal{P}}$ with $\mathbf{w}^*(\mathbf{x})$. More specifically, the shortage decreases at the rate $a_2 - d$ in $\bar{\mathcal{P}}$, while the average rate of decrease in shortage in $\mathcal{P}^\varepsilon$ is

$$\nu^1(\min\{m_1, m_2\} - d) + \nu^2(0 - d) + \nu^3(0 - d) + \nu^4(0 - d)$$
$$= \nu^1 \min\{m_1, m_2\} - d < a_2 - d.$$

Since c_2^- is strictly positive, the expected cost associated with $\mathbf{u}^*(\mathbf{x}, \mathbf{k})$ in $\mathcal{P}^\varepsilon$ is too large in comparison to the cost of $\mathbf{w}^*(\mathbf{x})$ in $\bar{\mathcal{P}}$. In other words, there is a significant loss of capacity on behalf of M_2 during the movement along $x_1 = 0, x_2 < 0$, whenever M_1 is under repair. This is a consequence of the requirement that $x_1(t)$ is not to become negative. It is important to emphasize that this is similar to the capacity loss phenomenon observed in Sethi and Zhou [18, Section 4], where we dealt with the case $a_1 < a_2$, i.e., when the maximum capacity of M_1 was strictly less than that of M_2. There we resorted to a switching manifold Γ (see Sethi and Zhou [18, Fig. 2]), below which we used a control policy that slowed down the rate of decrease in $x_1(t)$, while maintaining the same rate of decrease in shortage $(-x_2(t))$. This was accomplished by using full production capacities of both M_1 and M_2 between the switching manifold and the boundary $x_1 = 0$. This reduced the time spent on the boundary $x_1 = 0$, which mitigated the effect of the capacity loss phenomenon.

This insight obtained in [18, Section 4] suggests that a construction of an asymptotic optimal control for $\mathcal{P}^\varepsilon$ might require not too much time to be spent on the line $x_1 = 0, x_2 < 0$. One way to accomplish this would be to introduce a small region $\{\mathbf{x} | 0 \leq x_1 < \varepsilon^{\frac{1}{3}}, x_2 < 0\}$ as a neighborhood of $x_1 = 0, x_2 < 0$, where the policy is such that there is a tendency for the state trajectory to go away from $x_1 = 0$ and toward $x_1 = \varepsilon^{\frac{1}{3}}$, while still staying in the neighborhood. This tendency is in a marked contrast from the feedback policy $\mathbf{r}^*(\mathbf{x}, \mathbf{k})$, which brings the trajectory down to $x_1 = 0$ as quickly as possible.

Let us therefore introduce the following function:

$$\mathbf{r}^\varepsilon(\mathbf{x},\mathbf{k}) = \begin{cases} (0,0), & 0 \leq x_1 \leq B, x_2 > 0, \\ (0, \min\{k_2, d\}), & \varepsilon^{\frac{1}{3}} < x_1 \leq B, x_2 = 0, \\ (0, k_2), & \varepsilon^{\frac{1}{3}} < x_1 \leq B, x_2 < 0, \\ (\min\{k_1, k_2\}, k_2), & x_1 = \varepsilon^{\frac{1}{3}}, x_2 < 0, \\ (\min\{k_1, k_2, d\}, \min\{k_2, d\}), & x_1 = \varepsilon^{\frac{1}{3}}, x_2 = 0, \\ (k_1, k_2), & 0 < x_1 < \varepsilon^{\frac{1}{3}}, x_2 < 0, \\ (k_1, \min\{k_2, d\}), & 0 < x_1 < \varepsilon^{\frac{1}{3}}, x_2 = 0, \\ (k_1, \min\{k_1, k_2, d\}), & x_1 = x_2 = 0, \\ (k_1, \min\{k_1, k_2\}), & x_1 = 0, x_2 < 0. \end{cases} \quad (3.2)$$

It is interesting to point out that the line $x_1 = \varepsilon^{\frac{1}{3}}, x_2 < 0$ in defining (3.2) is a manifold introduced to reduce capacity losses on M_2 along $x_1 = 0, x_2 < 0$. Moreover, setting a policy $\mathbf{u}(\mathbf{x},\mathbf{k}) = \mathbf{r}^\varepsilon(\mathbf{x},\mathbf{k})$ implies that the full machine M_1 capacity be used below the manifold $x_1 = \varepsilon^{\frac{1}{3}}, x_2 < 0$. Let us now construct a control

$$\mathbf{w}^\varepsilon(\mathbf{x}) = \mathbf{r}^\varepsilon(\mathbf{x}, \mathbf{a}) \quad (3.3)$$

for $\bar{\mathcal{P}}$; see Fig.2. Note that $w_2^\varepsilon = w_2^*$ in the entire state space. Moreover, $w_1^\varepsilon = w_1^*$, except when $0 \leq x_1 \leq \varepsilon^{\frac{1}{3}}, x_2 \leq 0$. In this region of exception, the control policy has the tendency to go toward $(\varepsilon^{\frac{1}{3}}, 0)$. This policy, while not optimal, is nearly optimal for $\bar{\mathcal{P}}$ for sufficiently small ε, since $|x_1^\varepsilon - x_1^*| \leq \varepsilon^{\frac{1}{3}}$. More specifically, we have the following lemma which is easy to prove.

Lemma 3.1

There is a positive constant C such that

$$|J(\mathbf{x}, \mathbf{w}^\varepsilon) - v(\mathbf{x})| = |J(\mathbf{x}, \mathbf{w}^\varepsilon) - J(\mathbf{x}, \mathbf{w}^*)| \leq C\varepsilon^{\frac{1}{3}}.$$

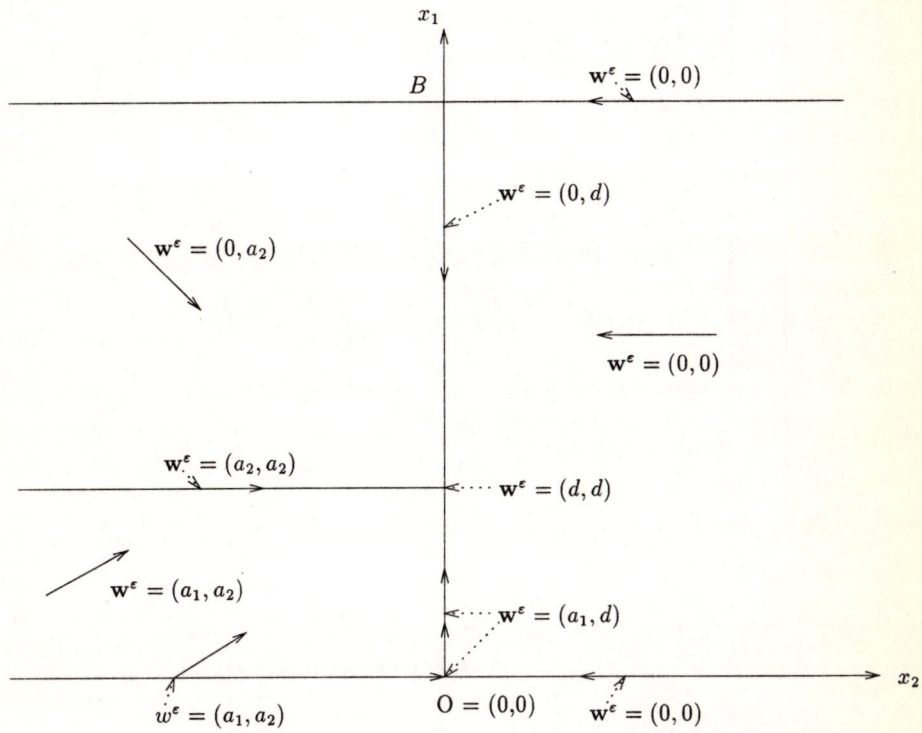

Figure 2: Near-Optimal Control $\mathbf{w}^\varepsilon$ for $\bar{\mathcal{P}}$ and Trajectory Movements

Note in Definition 2.3 that there are two alternative ways to define the class of admissible controls for the limiting problem. Control $\mathbf{w}^\varepsilon(\mathbf{x})$ constructed above can be equivalently written as

$$\mathbf{U}^\varepsilon(\mathbf{x}) = \begin{cases} (u_1^1(\mathbf{x}), u_1^2(\mathbf{x}), u_2^1(\mathbf{x}), u_2^3(\mathbf{x})) = \\ \quad (0,0,0,0), & 0 \leq x_1 \leq B, x_2 > 0, \\ \quad (0,0, \frac{d}{\nu^1+\nu^3}, \frac{d}{\nu^1+\nu^3}), & \varepsilon^{\frac{1}{3}} < x_1 \leq B, x_2 = 0, \\ \quad (0,0, m_2, m_2), & \varepsilon^{\frac{1}{3}} < x_1 \leq B, x_2 < 0, \\ \quad (\frac{\nu^1+\nu^3}{\nu^1+\nu^2} m_2, \frac{\nu^1+\nu^3}{\nu^1+\nu^2} m_2, m_2, m_2), & x_1 = \varepsilon^{\frac{1}{3}}, x_2 < 0, \\ \quad (\frac{d}{\nu^1+\nu^2}, \frac{d}{\nu^1+\nu^2}, \frac{d}{\nu^1+\nu^3}, \frac{d}{\nu^1+\nu^3}), & x_1 = \varepsilon^{\frac{1}{3}}, x_2 = 0, \\ \quad (m_1, m_1, m_2, m_2), & 0 < x_1 < \varepsilon^{\frac{1}{3}}, x_2 < 0, \\ \quad (m_1, m_1, \frac{d}{\nu^1+\nu^3}, \frac{d}{\nu^1+\nu^3}), & 0 < x_1 < \varepsilon^{\frac{1}{3}}, x_2 = 0, \\ \quad (m_1, m_1, \frac{d}{\nu^1+\nu^3}, \frac{d}{\nu^1+\nu^3}), & x_1 = x_2 = 0, \\ \quad (m_1, m_1, m_2, m_2), & x_1 = 0, x_2 < 0, \\ u_2^2(\mathbf{x}) = u_1^3(\mathbf{x}) = u_1^4(\mathbf{x}) = u_2^4(\mathbf{x}) = 0. \end{cases} \quad (3.4)$$

According to Lemma 3.1, the policy $\mathbf{U}^\varepsilon(\mathbf{x})$ is asymptotic optimal for $\bar{\mathcal{P}}$ as ε goes to zero.

In view of Lemma 3.1 and Theorem 2.2, a control for $\mathcal{P}^\varepsilon$, whose associated expected cost is close to $J(\mathbf{x}, \mathbf{U}^\varepsilon) = J(\mathbf{x}, \mathbf{w}^\varepsilon)$ would be asymptotic optimal. An obvious candidate is

$$\mathbf{u}^\varepsilon(\mathbf{x}, \mathbf{k}) = \mathbf{r}^\varepsilon(\mathbf{x}, \mathbf{k}). \quad (3.5)$$

In what follows, we shall prove its asymptotic optimality.

First, let us specify a notational convention. We denote

$$\mathbf{u}^\varepsilon(\mathbf{x},\mathbf{k}) = (u_1^\varepsilon(\mathbf{x},\mathbf{k}), u_2^\varepsilon(\mathbf{x},\mathbf{k})) = \begin{pmatrix} \mathbf{u}^\varepsilon(\mathbf{x},\mathbf{k}^1) \\ \mathbf{u}^\varepsilon(\mathbf{x},\mathbf{k}^2) \\ \mathbf{u}^\varepsilon(\mathbf{x},\mathbf{k}^3) \\ \mathbf{u}^\varepsilon(\mathbf{x},\mathbf{k}^4) \end{pmatrix} = \begin{pmatrix} u_1^\varepsilon(\mathbf{x},\mathbf{k}^1), u_2^\varepsilon(\mathbf{x},\mathbf{k}^1) \\ u_1^\varepsilon(\mathbf{x},\mathbf{k}^2), u_2^\varepsilon(\mathbf{x},\mathbf{k}^2) \\ u_1^\varepsilon(\mathbf{x},\mathbf{k}^3), u_2^\varepsilon(\mathbf{x},\mathbf{k}^3) \\ u_1^\varepsilon(\mathbf{x},\mathbf{k}^4), u_2^\varepsilon(\mathbf{x},\mathbf{k}^4) \end{pmatrix}.$$

Namely, each component of $\mathbf{u}^\varepsilon(\mathbf{x},\mathbf{k})$ represents the control policy under the corresponding machine state.

With this notation, the control $\mathbf{u}^\varepsilon(\mathbf{x},\mathbf{k})$ is shown in Fig. 3. We also show the state trajectory movements in Fig. 3 under each of the four machine states. In view of (3.4), (3.5) and (2.8), it is easy to see that the average movement of the state trajectory in Fig. 3 at each point of the state space coincides with the movement of the state trajectory of $\bar{\mathcal{P}}$ shown in Fig. 2 under the policy (3.3) or (3.4).

With this observation, we can expect the following main result of the paper to hold.

Theorem 3.1

There exists a positive constant C such that

$$|J^\varepsilon(\mathbf{x},\mathbf{k},\mathbf{u}^\varepsilon(\mathbf{x},\mathbf{k})) - J(\mathbf{x},\mathbf{U}^\varepsilon(\mathbf{x})))| < C\varepsilon^{\frac{1}{6}}, \qquad (3.6)$$

for sufficiently small ε.

Before we prove the theorem, we need to prove some preliminary results. For this purpose, we define the following regions in the state space:

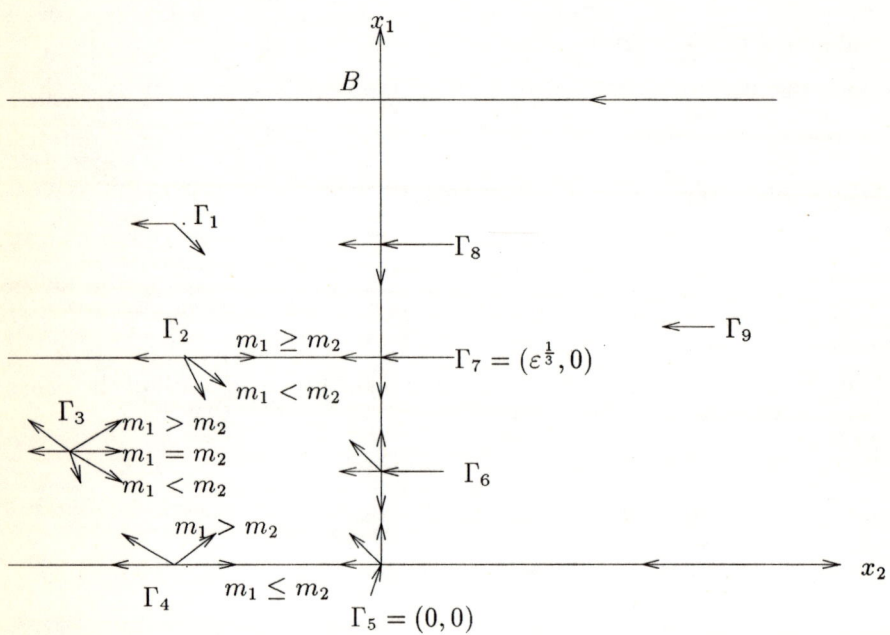

Figure 3: Asymptotic Optimal Control and Trajectory Movements under Different Machine States

$$\Gamma_1 = \{(x_1, x_2) | \varepsilon^{\frac{1}{3}} < x_1 \leq B, x_2 < 0\},$$

$$\Gamma_2 = \{(x_1, x_2) | x_1 = \varepsilon^{\frac{1}{3}}, x_2 < 0\},$$

$$\Gamma_3 = \{(x_1, x_2) | 0 < x_1 < \varepsilon^{\frac{1}{3}}, x_2 < 0\},$$

$$\Gamma_4 = \{(x_1, x_2) | x_1 = 0, x_2 < 0\},$$

$$\Gamma_5 = \{(x_1, x_2) | x_1 = 0, x_2 = 0\},$$

$$\Gamma_6 = \{(x_1, x_2) | 0 < x_1 < \varepsilon^{\frac{1}{3}}, x_2 = 0\},$$

$$\Gamma_7 = \{(x_1, x_2) | x_1 = \varepsilon^{\frac{1}{3}}, x_2 = 0\},$$

$$\Gamma_8 = \{(x_1, x_2) | \varepsilon^{\frac{1}{3}} < x_1 \leq B, x_2 = 0\},$$

$$\Gamma_9 = \{(x_1, x_2) | 0 \leq x_1 \leq B, x_2 > 0\}.$$

In connection with the above regions, $\Gamma_i, i = 1, 3, 9$ are called *interiors* and the remaining ones are called *boundaries*.

The main idea here is to show that the trajectories of the original problem and the limiting problem (under $\mathbf{u}^\varepsilon$ and $\mathbf{U}^\varepsilon$, respectively) are very close to each other in the average sense, if they start from the same initial state. It is very important to note that $\mathbf{u}^\varepsilon(\mathbf{x}, \mathbf{k}^i)$ coincides with $\mathbf{u}^i(\mathbf{y})$, the i-th component of $\mathbf{U}^\varepsilon(\mathbf{y})$, if $\mathbf{x}$ and $\mathbf{y}$ are in one of the same interiors.

Denote by $\mathbf{x}(\cdot)$ and $\mathbf{y}(\cdot)$, the aforementioned trajectories of the original and the limiting problems, respectively. Then we have

$$\frac{d}{dt}(x_1(t) - y_1(t))^2$$
$$= 2(x_1(t) - y_1(t))(\dot{x}_1(t) - \dot{y}_1(t))$$
$$= 2(x_1(t) - y_1(t))[(u_1^\varepsilon(\mathbf{x}(t), \mathbf{k}(\varepsilon, t)) - u_2^\varepsilon(\mathbf{x}(t), \mathbf{k}(\varepsilon, t)))$$
$$- \sum_{i=1}^{4} \nu^i(u_1^i(\mathbf{y}(t)) - u_2^i(\mathbf{y}(t)))] \qquad (3.7)$$
$$= 2(x_1(t) - y_1(t))\{\sum_{i=1}^{4} \chi_{\{\mathbf{k}(\varepsilon,t)=\mathbf{k}^i\}}[(u_1^\varepsilon(\mathbf{x}(t), \mathbf{k}^i) - u_1^i(\mathbf{y}(t)))$$
$$- (u_2^\varepsilon(\mathbf{x}(t), \mathbf{k}^i) - u_2^i(\mathbf{y}(t)))]$$
$$+ \sum_{i=1}^{4} (\chi_{\{\mathbf{k}(\varepsilon,t)=\mathbf{k}^i\}} - \nu^i)(u_1^i(\mathbf{y}(t)) - u_2^i(\mathbf{y}(t)))\}.$$

Define

$$\lambda^i(\mathbf{x}, \mathbf{y}) = (x_1 - y_1)[(u_1^\varepsilon(\mathbf{x}, \mathbf{k}^i) - u_1^i(\mathbf{y})) - (u_2^\varepsilon(\mathbf{x}, \mathbf{k}^i) - u_2^i(\mathbf{y}))], i = 1, 2, 3, 4.$$

In the following, we want to show that

$$E \int_0^t \lambda^i(\mathbf{x}(s), \mathbf{y}(s)) \le K\varepsilon^{\frac{1}{3}}(1+t), i = 1, 2, 3, 4, \qquad (3.8)$$

for some constant K. To this end, let us first introduce several useful lemmas.

Lemma 3.2

For $i = 1, 2, 3, 4$, we have

(a) $\lambda^i(\mathbf{x}, \mathbf{y}) = 0$, for $(\mathbf{x}, \mathbf{y}) \in \Gamma_1 \times \Gamma_1$;

(b) $\lambda^i(\mathbf{x}, \mathbf{y}) \le 0$, for $(\mathbf{x}, \mathbf{y}) \in [\Gamma_1 \times (\Gamma_2 \cup \Gamma_7)] \cup [(\Gamma_2 \cup \Gamma_7) \times \Gamma_1]$;

(c) $\lambda^i(\mathbf{x}, \mathbf{y}) \le 0$, for $\mathbf{x} \in \{(x_1, x_2) | 0 \le x_1 \le B, x_2 < 0\}, \mathbf{y} \in \Gamma_7$;

(d) $\lambda^i(\mathbf{x}, \mathbf{y}) \le K\varepsilon^{\frac{1}{3}}$, for $\mathbf{x}, \mathbf{y} \in \{(x_1, x_2) | 0 \le x_1 \le \varepsilon^{\frac{1}{3}}\}$.

Proof

(a) Since Γ_1 is an interior in which $u_j^\varepsilon(\mathbf{x}, \mathbf{k}^i) = u_j^i(\mathbf{y})$ for $i = 1, 2, 3, 4$ and $j = 1, 2$, we have $\lambda^i(\mathbf{x}, \mathbf{y}) = 0$ on $\Gamma_1 \times \Gamma_1$ by definition.

(b) This can be proved by direct calculation in each case. For example, when $\mathbf{x} \in \Gamma_1, \mathbf{y} \in \Gamma_2$, we have

$x_1 - y_1 > 0$, and

$$(u_1^\varepsilon(\mathbf{x}, \mathbf{k}^1) - u_1^1(\mathbf{y})) - (u_2^\varepsilon(\mathbf{x}, \mathbf{k}^1) - u_2^1(\mathbf{y})) = (0 - \frac{\nu^1+\nu^3}{\nu^1+\nu^2}m_2) - (m_2 - m_2) \leq 0,$$

which gives $\lambda^1(\mathbf{x}, \mathbf{y}) \leq 0$.

(c) This can also be shown by direct calculation.

(d) If both $\mathbf{x}$ and $\mathbf{y}$ are in the ε-strip $\{(x_1, x_2) | 0 \leq x_1 \leq \varepsilon^{\frac{1}{3}}\}$, then $|x_1 - y_1| \leq \varepsilon^{\frac{1}{3}}$. The desired result is then easily seen, since all the controls constructed are bounded (independent of ε).

Lemma 3.3

Let $\tilde{\mathbf{x}}(\cdot)$ be the trajectory of $\mathcal{P}^\varepsilon$ under $\mathbf{u}^\varepsilon(\mathbf{x}, \mathbf{k})$, and $\tilde{\mathbf{y}}(\cdot)$ be the trajectory of $\bar{\mathcal{P}}$ under $\mathbf{U}^\varepsilon(\mathbf{x})$, both starting from the same initial $\tilde{\mathbf{x}} \in \Gamma_8$. Then

$$\tilde{x}_1(t) \geq \tilde{y}_1(t), \quad \text{for } t \in [0, \frac{\tilde{x}_1 - \varepsilon^{\frac{1}{3}}}{d}]. \tag{3.9}$$

Proof

Let us pick up a sample point ω. Suppose during the time interval $[0, \tau)$, the machine state is in $\{\mathbf{k}^2, \mathbf{k}^4\}$, and then switch to be in $\{\mathbf{k}^1, \mathbf{k}^3\}$ (τ could be zero) at τ. Then obviously (3.9) holds for $t \in [0, \tau]$. Note that for any $t \in [\tau, \frac{\tau m_2}{m_2-d}]$, we have

$$\tilde{x}_1(t) = \tilde{x}_1 - (t-\tau)m_2 \geq \tilde{x}_1 - td = \tilde{y}_1(t),$$

and the equality holds only at $t = t_0 \equiv \frac{\tau m_2}{m_2-d}$.

Here t_0 is nothing but the time of $\tilde{\mathbf{x}}(\cdot)$ hitting Γ_8 under the assumption that the machine state does not switch to $\{\mathbf{k}^2, \mathbf{k}^4\}$ in $[\tau, t_0)$. Now let us assume that the machine state changes to $\{\mathbf{k}^2, \mathbf{k}^3\}$ at time $\theta < t_0$ and stays there until time τ'. By the same argument as above, it is easy to check that $\tilde{x}_1(t) \geq \tilde{y}_1(t)$ for $t \geq \theta$. This situation continues until $t = \frac{\tilde{x}_1 - \varepsilon^{\frac{1}{3}}}{d}$, the time of $\tilde{\mathbf{y}}(\cdot)$ reaching the threshold Γ_7.

Lemma 3.4

Let $\tilde{\mathbf{x}}(\cdot)$ be the trajectory of $\mathcal{P}^\varepsilon$ under $\mathbf{u}^\varepsilon(\mathbf{x}, \mathbf{k})$ with initial $\tilde{\mathbf{x}} \in \Gamma_1 \cup \Gamma_8$, and $\tilde{\mathbf{y}}(\cdot)$ be the trajectory of $\bar{\mathcal{P}}$ under $\mathbf{U}^\varepsilon(\mathbf{x})$ with initial $\tilde{\mathbf{y}} \in \Gamma_8$. If $|\tilde{\mathbf{x}} - \tilde{\mathbf{y}}| \leq C \varepsilon^{\frac{1}{3}}$, then

$$\tilde{x}_1(t) \geq \tilde{y}_1(t) - C \frac{m_2}{m_2 - d} \varepsilon^{\frac{1}{3}}, \text{ for } t \in [0, \frac{\tilde{y}_1 - \varepsilon^{\frac{1}{3}}}{d}]. \quad (3.10)$$

Proof

Define

$$\mathbf{x}'(t) = (x_1'(t), x_2'(t)) = \begin{cases} (\tilde{x}_1, \tilde{y}_2 - (m_2 - d)t), & 0 \leq t < \frac{\tilde{y}_2 - \tilde{x}_2}{m_2 - d} \\ (\tilde{x}_1(t - \frac{\tilde{y}_2 - \tilde{x}_2}{m_2 - d}), \tilde{x}_2(t - \frac{\tilde{y}_2 - \tilde{x}_2}{m_2 - d})), & t \geq \frac{\tilde{y}_2 - \tilde{x}_2}{m_2 - d}, \end{cases}$$

$$\mathbf{y}'(t) = (y_1'(t), y_2'(t)) = (\tilde{y}_1(t) - \tilde{y}_1 + \tilde{x}_1, \tilde{y}_2(t)).$$

Then $\mathbf{x}'(0) = \mathbf{y}'(0) = (\tilde{x}_1, \tilde{y}_2) \in \Gamma_8$. We can therefore apply Lemma 3.3 to obtain

$$\begin{aligned}
\tilde{x}_1(t) &= x_1'(t + \tfrac{\tilde{y}_2 - \tilde{x}_2}{m_2 - d}) \geq y_1'(t + \tfrac{\tilde{y}_2 - \tilde{x}_2}{m_2 - d}) \\
&= \tilde{y}_1(t + \tfrac{\tilde{y}_2 - \tilde{x}_2}{m_2 - d}) - \tilde{y}_1 + \tilde{x}_1 = \tilde{y}_1(t) - d \tfrac{\tilde{y}_2 - \tilde{x}_2}{m_2 - d} - \tilde{y}_1 + \tilde{x}_1 \\
&\geq \tilde{y}_1(t) - C \tfrac{m_2}{m_2 - d} \varepsilon^{\frac{1}{3}},
\end{aligned}$$

for $t \in [0, \frac{\tilde{y}_1 - \varepsilon^{\frac{1}{3}}}{d}]$. This proves the lemma.

Let us now turn to the proof of the main result.

Proof of Theorem 3.1

We have to analyze various different cases depending on the position of the initial state $\mathbf{x}$.

Case I. The initial state $\mathbf{x} \in \Gamma_1$.

In this case, both trajectories have the following important features:

(i) The deterministic trajectory $\mathbf{y}(\cdot)$ will either first hit Γ_7 and then stay there forever, or first hit either Γ_2 or Γ_8 and then go along one of these boundaries towards Γ_7.

(ii) The stochastic trajectory $\mathbf{x}(\cdot)$ will never go up inside Γ_1, and once having entered the "box" $\{(x_1, x_2) | 0 \leq x_1 \leq \varepsilon, x_2 \leq 0\}$, it will never go out of it.

Define by τ_x and τ_y, respectively, the stopping times of $\mathbf{x}(\cdot)$ and $\mathbf{y}(\cdot)$ hitting the boundary $\Gamma_2 \cup \Gamma_7 \cup \Gamma_8$. Note that τ_y is deterministic. We now want to prove the following estimate:

$$P(\tau_y - \tau_x \geq \varepsilon^{\frac{1}{3}}) \leq C\varepsilon, \qquad (3.11)$$

for some positive constant C.

To show this, we only consider the case when $\mathbf{y}(\cdot)$ hits Γ_8. The analyses for the other cases are the same. By Corollary 2.1,

$$P(|\mathbf{x}(\tau_x) - \mathbf{y}(\tau_x)| \geq (a_2 - d)\varepsilon^{\frac{1}{3}}; \tau_x \leq \tau_y)$$
$$\leq P(|\int_0^{\tau_x} \sum_{i=0}^{4}(\chi_{\{\mathbf{k}(\varepsilon,t) = \mathbf{k}^i\}} - \nu^i)(u_1^i(\mathbf{y}) - u_2^i(\mathbf{y}))dt| > \tfrac{1}{2}(a_2 - d)\varepsilon^{\frac{1}{3}}; \tau_x \leq \tau_y)$$
$$+ P(|\int_0^{\tau_x} \sum_{i=0}^{4}(\chi_{\{\mathbf{k}(\varepsilon,t) = \mathbf{k}^i\}} - \nu^i)(u_2^i(\mathbf{y}) - d)dt| > \tfrac{1}{2}(a_2 - d)\varepsilon^{\frac{1}{3}}; \tau_x \leq \tau_y)$$
$$\leq C(e^{-K\varepsilon^{-1}} + e^{-K\varepsilon^{-1/3}(1+\tau_y)^{-3}}) \leq C\varepsilon.$$

Hence,

$$P(\tau_y - \tau_x \geq \varepsilon^{\frac{1}{3}})$$
$$\leq P(|\mathbf{x}(\tau_x) - \mathbf{y}(\tau_x)| \geq (a_2 - d)\varepsilon^{\frac{1}{3}}; \tau_x \leq \tau_y; \mathbf{x}(\tau_x) \in \Gamma_8 \cup \Gamma_7)$$
$$+ P(|\mathbf{x}(\tau_x) - \mathbf{y}(\tau_x)| \geq y_1(\tau_y); \tau_x \leq \tau_y; \mathbf{x}(\tau_x) \in \Gamma_2)$$
$$\leq C\varepsilon.$$

This proves (3.11).

To complete the proof of the theorem in Case I, we have to study three subcases:

Case I.1. $\mathbf{y}(\tau_y) \in \Gamma_2$.

In this case, we note the following facts:

(i) $\lambda^i(\mathbf{x}(t), \mathbf{y}(t)) = 0$, $t < \min\{\tau_x, \tau_y\}$, by virtue of Lemma 3.2 (a).

(ii) If $\tau_y \leq \tau_x$, then for $t \geq \tau_y$, we have $\mathbf{y}(t) \in \Gamma_2 \cup \Gamma_7$ and $\mathbf{x}(t) \in \cup_{i=1}^{7}\Gamma_i$. Hence, by Lemma 3.2 (b) and (c), we have

$$\lambda^i(\mathbf{x}(t), \mathbf{y}(t)) \leq K\varepsilon^{\frac{1}{3}}, i = 1, 2, 3, 4.$$

(iii) If $\tau_y > \tau_x$, then as $t \geq \tau_x$, $\mathbf{y}(t) \in \Gamma_1 \cup \Gamma_2 \cup \Gamma_7$ and $\mathbf{x}(t) \in \cup_{i=2}^{7}\Gamma_i$. Again by Lemma 3.2 (b) and (c), we have

$$\lambda^i(\mathbf{x}(t), \mathbf{y}(t)) \leq K\varepsilon^{\frac{1}{3}}, i = 1, 2, 3, 4.$$

For any $0 \leq t \leq \tau_y$, we can now estimate

$E \int_0^t \lambda^i(\mathbf{x}(s), \mathbf{y}(s)) ds$

$= E(\int_0^t \lambda^i(\mathbf{x}(s), \mathbf{y}(s)) ds; \tau_y \leq \tau_x) + E(\int_0^t \lambda^i(\mathbf{x}(s), \mathbf{y}(s)) ds; \tau_y > \tau_x)$

$= E(\int_0^t \lambda^i(\mathbf{x}(s), \mathbf{y}(s)) ds; \tau_y > \tau_x)$

$\leq E(\int_0^t \lambda^i(\mathbf{x}(s), \mathbf{y}(s)) ds; \tau_x < \tau_y < \tau_x + \varepsilon^{\frac{1}{3}}) + KP(\tau_y - \tau_x \geq \varepsilon^{\frac{1}{3}})$

$= E(\int_{\min\{t,\tau_x\}}^t \lambda^i(\mathbf{x}(s), \mathbf{y}(s)) ds; \tau_x < \tau_y < \tau_x + \varepsilon^{\frac{1}{3}}) + KP(\tau_y - \tau_x \geq \varepsilon^{\frac{1}{3}})$

$\leq K\varepsilon^{\frac{1}{3}} + C\varepsilon.$

For any $t > \tau_y$, by (ii) and (iii) above,

$E(\int_{\tau_y}^t \lambda^i(\mathbf{x}(s), \mathbf{y}(s)) ds)$

$= E(\int_{\tau_y}^t \lambda^i(\mathbf{x}(s), \mathbf{y}(s)) ds; \tau_y \leq \tau_x) + E(\int_{\tau_y}^t \lambda^i(\mathbf{x}(s), \mathbf{y}(s)) ds; \tau_y > \tau_x)$

$\leq K\varepsilon^{\frac{1}{3}}t.$

Therefore, (3.8) follows in Case I.

Case I.2. $\mathbf{y}(\tau_y) \in \Gamma_8$.

In this case, let us first estimate

$P(|\mathbf{x}(\tau_y) - \mathbf{y}(\tau_y)| > \varepsilon^{\frac{1}{3}}; \tau_y \leq \tau_x)$

$\leq P(|\int_0^{\tau_y} \sum_{i=0}^{4}(\chi_{\{k(\varepsilon,t)=k^i\}} - \nu^i)(u_1^i(\mathbf{y}) - u_2^i(\mathbf{y}))dt| > \frac{1}{2}\varepsilon^{\frac{1}{3}})$

$+ P(|\int_0^{\tau_y} \sum_{i=0}^{4}(\chi_{\{k(\varepsilon,t)=k^i\}} - \nu^i)(u_2^i(\mathbf{y}) - d)dt| > \frac{1}{2}\varepsilon^{\frac{1}{3}})$

$\leq C\varepsilon.$

Furthermore, with $\gamma = 2m_2 - d$ and $\bar{\gamma} = 2a_2 - d$, we have

$$P(|\mathbf{x}(\tau_y) - \mathbf{y}(\tau_y)| > (1 + \gamma + \bar{\gamma})\varepsilon^{\frac{1}{3}}; \tau_y > \tau_x)$$
$$= P(|\mathbf{x}(\tau_y) - \mathbf{y}(\tau_y)| > (1 + \gamma + \bar{\gamma})\varepsilon^{\frac{1}{3}}; \tau_x < \tau_y < \tau_x + \varepsilon^{\frac{1}{3}}) + P(\tau_y \geq \tau_x + \varepsilon^{\frac{1}{3}})$$
$$\leq P(|\mathbf{x}(\tau_y) - \mathbf{x}(\tau_x)| > \gamma\varepsilon^{\frac{1}{3}}; \tau_x < \tau_y < \tau_x + \varepsilon^{\frac{1}{3}})$$
$$+ P(|\mathbf{x}(\tau_x) - \mathbf{y}(\tau_x)| > \varepsilon^{\frac{1}{3}}; \tau_x < \tau_y < \tau_x + \varepsilon^{\frac{1}{3}})$$
$$+ P(|\mathbf{y}(\tau_x) - \mathbf{y}(\tau_y)| > \bar{\gamma}\varepsilon^{\frac{1}{3}}; \tau_x < \tau_y < \tau_x + \varepsilon^{\frac{1}{3}}) + C\varepsilon$$
$$= P(|\mathbf{x}(\tau_x) - \mathbf{y}(\tau_x)| > \varepsilon^{\frac{1}{3}}; \tau_x < \tau_y < \tau_x + \varepsilon^{\frac{1}{3}}) + C\varepsilon$$
$$\leq C'\varepsilon.$$

Combining the above two estimates, we conclude that

$$P(|\mathbf{x}(\tau_y) - \mathbf{y}(\tau_y)| > (1 + \gamma + \bar{\gamma})\varepsilon^{\frac{1}{3}}) \leq C\varepsilon.$$

Observe the following facts:

(i) If for a sample point ω, $|\mathbf{x}(\tau_y) - \mathbf{y}(\tau_y)| \leq (1 + \gamma + \bar{\gamma})\varepsilon^{\frac{1}{3}}$, then by Lemma 3.4,

$$x_1(t) \geq y_1(t) - \frac{(1 + \gamma + \bar{\gamma})k_2}{k_2 - d}\varepsilon^{\frac{1}{3}}, \ t \in [\tau_y, \tau_y + \frac{y_1(\tau_y) - \varepsilon^{\frac{1}{3}}}{d}].$$

Hence,

$$\lambda^i(\mathbf{x}(t), \mathbf{y}(t)) \leq K\varepsilon^{\frac{1}{3}}, \ t \in [\tau_y, \tau_y + \frac{y_1(\tau_y) - \varepsilon^{\frac{1}{3}}}{d}].$$

(ii) When $t \geq \frac{y_1(\tau_y) - \varepsilon^{\frac{1}{3}}}{d}$, $y(t) \in \Gamma_7$. In this case, we have by Lemma 3.2 (c) that

$$\lambda^i(\mathbf{x}(t), \mathbf{y}(t)) \leq 0.$$

For any $0 \leq t \leq \tau_y$, we have the following, similar to Case I.1:

$$E \int_0^t \lambda^i(\mathbf{x}(s), \mathbf{y}(s))ds \leq K\varepsilon^{\frac{1}{3}} + C\varepsilon.$$

On the other hand, for any $t \geq \tau_y$, we obtain

$$E \int_{\tau_y}^{t} \lambda^i(\mathbf{x}(s), \mathbf{y}(s)) ds$$

$$\leq E \int_{\tau_y}^{\frac{y_1(\tau_y) - \varepsilon^{\frac{1}{3}}}{d}} \lambda^i(\mathbf{x}(s), \mathbf{y}(s)) ds$$

$$= E(\int_{\tau_y}^{\frac{y_1(\tau_y) - \varepsilon^{\frac{1}{3}}}{d}} \lambda^i(\mathbf{x}(s), \mathbf{y}(s)) ds; |\mathbf{x}(\tau_y) - \mathbf{y}(\tau_y)| \leq (1 + \gamma + \bar{\gamma})\varepsilon^{\frac{1}{3}})$$

$$+ KP(|\mathbf{x}(\tau_y) - \mathbf{y}(\tau_y)| > (1 + \gamma + \bar{\gamma})\varepsilon^{\frac{1}{3}})$$

$$\leq K\varepsilon^{\frac{1}{3}} + K\varepsilon.$$

This proves (3.8) in Case I.2.

Case I.3. $\mathbf{y}(\tau_y) \in \Gamma_7$.

In this case, (3.8) follows easily from Lemma 3.2 (c).

Case II. The initial state $\mathbf{x} \in \cup_{i=2}^{7} \Gamma_i$.

In this case, both $\mathbf{x}(\cdot)$ and $\mathbf{y}(\cdot)$ will never go out of the ε-strip $\{(x_1, x_2) | 0 \leq x_1 \leq \varepsilon^{\frac{1}{3}}\}$. Therefore, (3.8) holds by Lemma 3.2 (d).

Case III. The initial state $\mathbf{x} \in \Gamma_9 \cup \Gamma_8$.

If $\mathbf{x} \in \Gamma_9$, then initially both $\mathbf{x}(\cdot)$ and $\mathbf{y}(\cdot)$ will coincide until they hit $\cup_{i=5}^{8} \Gamma_i$. So we may assume that $\mathbf{x} \in \Gamma_8$.

By Lemma 3.3, if $t \leq \frac{x_1 - \varepsilon^{\frac{1}{3}}}{d}$, it holds that $x_1(t) \geq y_1(t)$, and therefore $\lambda^i(\mathbf{x}(t), \mathbf{y}(t)) \leq 0$. When $t \geq \frac{x_1 - \varepsilon^{\frac{1}{3}}}{d}$, $y(t) \in \Gamma_7$. In this case, $\lambda^i(\mathbf{x}(t), \mathbf{y}(t)) \leq 0$ as given by Lemma 3.2 (c). So (3.8) holds.

We have now proved (3.8) for all of the cases. By (3.7), (3.8), and Lemma 2.1, we obtain

$$E(x_1(t) - y_1(t))^2 \leq 8K(1 + t)\varepsilon^{\frac{1}{3}} + \frac{1}{2} E(x_1(t) - y_1(t))^2$$

$$+ \frac{1}{2} E |\int_0^t \sum_{i=1}^{4} (\chi_{\{\mathbf{k}(\varepsilon, t) = \mathbf{k}^i\}} - \nu^i)(u_1^i(\mathbf{y}(t)) - u_2^i(\mathbf{y}(t))) dt|^2$$

$$\leq 8K(1 + t)\varepsilon^{\frac{1}{3}} + \frac{1}{2} E(x_1(t) - y_1(t))^2 + C(1 + t^2)^{\frac{1}{2}} \varepsilon^{\frac{1}{2}}$$

$$\leq C(1 + t)\varepsilon^{\frac{1}{3}} + \frac{1}{2} E(x_1(t) - y_1(t))^2.$$

Thus,

$$E|x_1(t) - y_1(t)| \leq C(1 + t)^{\frac{1}{2}} \varepsilon^{\frac{1}{6}}.$$

By a similar argument, we can get

$$E|x_2(t) - y_2(t)| \leq C(1+t)^{\frac{1}{2}}\varepsilon^{\frac{1}{6}}.$$

This concludes the proof.

Theorem 3.2

There exists a positive constant C such that

$$|J^\varepsilon(\mathbf{x}, \mathbf{k}, \mathbf{u}(\mathbf{x})) - v^\varepsilon(\mathbf{x}, \mathbf{k})| < K_5 \varepsilon^{\frac{1}{6}}. \qquad (3.12)$$

Proof

This follows easily from Theorems 2.2, 3.1 and Lemma 3.1.

3.2 The Case $c_1 \geq c_2^+$

We write the optimal control obtained in (2.10) as $\mathbf{w}^*(\mathbf{x}) = \mathbf{r}^*(\mathbf{x}, \mathbf{a})$, where

$$\mathbf{r}^*(\mathbf{x}, \mathbf{k}) = \begin{cases} (\min\{k_1, k_2, d\}, \min\{k_1, k_2, d\}), & x_1 = x_2 = 0, \\ (0, k_2), & 0 < x_1 \leq B, \\ (0, 0), & x_1 = 0, x_2 > 0, \\ (\min\{k_1, k_2\}, \min\{k_1, k_2\}), & x_1 = 0, x_2 < 0. \end{cases} \qquad (3.13)$$

Again we introduce a switching manifold slightly above the boundary $x_1 = 0, x_2 < 0$ as in Section 3.1, and define the following function:

$$\mathbf{r}^\varepsilon(\mathbf{x},\mathbf{k}) = \begin{cases} (0,k_2), & 0 \leq x_1 \leq B, x_2 > 0, \\ (0,k_2), & \varepsilon^{\frac{1}{3}} < x_1 \leq B, x_2 \leq 0, \\ (\min\{k_1,k_2\},k_2), & x_1 = \varepsilon^{\frac{1}{3}}, x_2 < 0, \\ (\min\{k_1,k_2,d\},\min\{k_2,d\}), & x_1 = \varepsilon^{\frac{1}{3}}, x_2 = 0, \\ (k_1,k_2), & 0 < x_1 < \varepsilon^{\frac{1}{3}}, x_2 < 0, \\ (k_1,\min\{k_2,d\}), & 0 < x_1 < \varepsilon^{\frac{1}{3}}, x_2 = 0, \\ (k_1,\min\{k_1,k_2,d\}), & x_1 = x_2 = 0, \\ (k_1,\min\{k_1,k_2\}), & x_1 = 0, x_2 < 0, \end{cases} \quad (3.14)$$

Finally, we construct a control

$$\mathbf{u}^\varepsilon(\mathbf{x},\mathbf{k}) = \mathbf{r}^\varepsilon(\mathbf{x},\mathbf{k}), \quad (3.15)$$

which can be shown to be asymptotic optimal for $\mathcal{P}^\varepsilon$ in a manner similar to the proof of Theorem 3.1.

4 Analysis of Kanban Control

Kanban control is a threshold type control. It is defined as follows for some $\{\theta_1(\varepsilon),\theta_2(\varepsilon)\}$ with $\theta_1(\varepsilon) \geq 0$ and $\theta_2(\varepsilon) \geq 0$.

$$\mathbf{u}_K(\mathbf{x}, \mathbf{k}) = \begin{cases} (0,0), & \theta_1(\varepsilon) \le x_1 \le B, x_2 > \theta_2(\varepsilon), \\ (0, \min\{k_2, d\}), & \theta_1(\varepsilon) < x_1 \le B, x_2 = \theta_2(\varepsilon), \\ (0, k_2), & \theta_1(\varepsilon) < x_1 \le B, x_2 < \theta_2(\varepsilon), \\ (\min\{k_1, k_2\}, k_2), & x_1 = \theta_1(\varepsilon), x_2 < \theta_2(\varepsilon), \\ (\min\{k_1, k_2, d\}, \min\{k_2, d\}), & x_1 = \theta_1(\varepsilon), x_2 = \theta_2(\varepsilon), \\ (k_1, k_2), & 0 < x_1 < \theta_1(\varepsilon), x_2 < \theta_2(\varepsilon), \\ (k_1, \min\{k_2, d\}), & 0 \le x_1 < \theta_1(\varepsilon), x_2 = \theta_2(\varepsilon), \\ (k_1, \min\{k_1, k_2, d\}), & x_1 = 0, x_2 = \theta_2(\varepsilon), \\ (k_1, \min\{k_1, k_2\}), & x_1 = 0, x_2 < \theta_2(\varepsilon), \\ (k_1, 0), & 0 \le x_1 < \theta_1(\varepsilon), x_2 > \theta_2(\varepsilon). \end{cases}$$
(4.1)

The Kanban Control (KC) is an adaptation of Just-In-Time (JIT) method to our stochastic problem. Indeed Kanban control reduces to conventional JIT when $\theta_1(\varepsilon) = \theta_2(\varepsilon) = 0$. But given unreliable machines, one can lower the cost by selecting nonnegative values for the threshold inventory levels $\theta_1(\varepsilon)$ and $\theta_2(\varepsilon)$. This is because positive inventory levels hedge against machine breakdowns. While the general idea seems to have been around, the formula (4.1) appears in Sethi *et al.* [13] for the first time. It should be noted that under Assumptions (A1) - (A2), KC is also asymptotically optimal.

We can state and prove the following results.

Theorem 4.1

Kanban control is asymptotically optimal for $a_1 \ge a_2$ and $c_1 \le c_2^+$, provided $\theta_1(\varepsilon) = C_1 \varepsilon^{\frac{1}{3}}$ and $\theta_2(\varepsilon) = C_2 \varepsilon^{\frac{1}{3}}$ for some constants $C_1 > 0$ and $C_2 \ge 0$.

Proof

First we note that Theorem 3.2 could be proved in the same manner if we had replaced the switching manifold $x_2 = 0$ in Fig. 3 by the manifold $x_2 = \theta_2(\varepsilon)$. With this modification, the only difference between the control in Fig. 3 and the Kanban control is in the region $x_2 > \theta_2(\varepsilon)$ and $0 \leq x_1 < \theta_1(\varepsilon)$. However, if the initial state **x** is in this region, the cost of Kanban control and the cost of hierarchical control cannot differ by more than $C\varepsilon^{\frac{1}{3}}$, where $C > 0$ is some constant.

Theorem 4.2

Kanban control is not asymptotically optimal for $a_1 > a_2$ and $c_1 > c_2^+$ even if $\theta_1(\varepsilon) = C_1 \varepsilon^{\frac{1}{3}}$ and $\theta_2(\varepsilon) = C_2 \varepsilon^{\frac{1}{3}}$ for some $C_1 > 0$ and $C_2 \geq 0$.

Proof

In this case, it is obvious that the trajectories of Kanban control will not converge to the trajectories of Fig. 3 in Sethi and Zhou [18]. Clearly, the Kanban control cannot be asymptotically optimal.

5 Concluding Remarks

In this paper we have constructed asymptotic optimal feedback controls for a stochastic two-machine flowshop for the first time. We introduce a switching manifold (in the case when the average capacity of the first machine exceeds that of the second), whose purpose it is to increase the inventory in the internal buffer when it is too close to being empty. This reduced the amount of time during which the internal buffer is empty and during which the second machine cannot be used, even when it is up, if the first machine is down.

This paper considers only the case when the internal buffer is of finite size. Fong and Zhou [4] have studied this problem also with a finite external buffer. They have derived hierarchical open-loop controls by developing two techniques called "weak-Lipschitz property" and "state domain approximation". Recently, Fong and Zhou [5] have used the "state domain approximation" approach also to obtain hierarchical feedback controls for the problem.

We conclude this paper by mentioning that construction of asymptotic optimal feedback controls for general dynamic stochastic flowshops and jobshops remains a wide open research area. We are able to treat a simple two-machine flowshop, since we were able to explicitly solve the limiting problem. What one would like to have is a general methodology that can provably construct asymptotic optimal feedback controls, without requiring a detailed characterization of the optimal feedback controls of the limiting problems. It should be indicated that such a methodology is available for open loop controls; see Sethi, Zhang and Zhou [15, 19]. This methodology, which is based on the Lipschitz property of open loop controls, cannot be extended, however, to feedback controls that are not likely to be Lipschitz.

References

[1] R. Akella, Y.F. Choong, and S.B. Gershwin, Performance of hierarchical production scheduling policy, *IEEE Trans. Components, Hybrids, and Manufacturing Technology,* **CHMT-7** (1984).

[2] R. Akella and P.R. Kumar, Optimal control of production rate in failure prone manufacturing system, *IEEE Trans. Automat. Control* **AC-31** (1986), 116-126.

[3] S. Bai and S.B. Gershwin, Scheduling manufacturing systems with work-in-process inventory, *Proc. of the 29th IEEE Conference on Decision and Control,* Honolulu, HI (1990), 557-564.

[4] N.T. Fong and X.Y. Zhou, Hierarchical production policies in stochastic two-machine flowshops with finite buffers, to appear in *Journal of Optimization Theory and Applications.*

[5] N.T. Fong and X.Y. Zhou, Feedback controls of stochastic two-machine flowshops with state constraints, Working paper.

[6] S.B. Gershwin, R. Akella, and Y.F. Choong, Short-term production scheduling of an automated manufacturing facility, *IBM Journal of Research and Development*, **29** (1985), 392-400.

[7] N. Ikeda and S. Watanabe, *Stochastic Differential Equations and Diffusion Processes*, 2nd ed., Kodansha/North-Holland, Tokyo, 1989.

[8] S.X.C. Lou and G. van Ryzin, Optimal control rules for scheduling job shops, *Annals of Operations Research*, **17** (1989), 233-248.

[9] S.X.C. Lou, S. Sethi, and Q. Zhang, Optimal feedback production planning in a stochastic two–machine flowshop, *European Journal of Operational Research-Special Issue on Stochastic Control Theory and Operational Research,* **73**, 331- 345.

[10] E. Presman, S. P. Sethi, and Q. Zhang, Optimal feedback production planning in a stochastic N−machine flowshop, to appear in *Proc. of 12th World Congress of International Federation of Automatic Control*, Sydney, Australia (1993); full paper to appear in *Automatica*.

[11] G. van Ryzin, S.X.C. Lou, and S. B. Gershwin, Production control for a tandem two–machine system, *IIE Transactions*, **25**(1993), 5-20.

[12] C. Samaratunga, S.P. Sethi, and X. Y. Zhou, Computational evaluation of hierarchical production control policies for stochastic manufacturing systems, to appear in *Operations Research*.

[13] S.P. Sethi, H. Yan, Q. Zhang, and X.Y. Zhou, Feedback production planning in a stochastic two machine flow shop: Asymptotic analysis and computational results, *International Journal of Production Economics*, **30-31**(1993), 79-93.

[14] S.P. Sethi and Q. Zhang, *Hierarchical Decision Making in Stochastic Manufacturing Systems*, Birkhäuser Boston, Cambridge, MA, 1994.

[15] S.P. Sethi, Q. Zhang, and X.Y. Zhou, Hierarchical control for manufacturing systems with machines in tandem. *Stochastics and Stochastics Reports*, **41** (1992), 89-118.

[16] S. P. Sethi, Q. Zhang, and X. Y. Zhou, Hierarchical controls in stochastic manufacturing systems with convex costs, *Journal of Optimization Theory and Applications,* **80** (1994), 299-317.

[17] S. P. Sethi, Q. Zhang, and X. Y. Zhou, Hierarchical production controls in a stochastic two-machine flowshop with a finite internal buffer, Working Paper.

[18] S. P. Sethi and X. Y. Zhou, Optimal feedback controls in deterministic dynamic two-machine flowshops, Working Paper.

[19] S. P. Sethi and X. Y. Zhou, Stochastic dynamic jobshops and hierarchical production planning, *IEEE Transactions on Automatic Control,* **AC-39**, (1994), 2061-2076.

[20] N. Srivatsan, S. Bai, and S. Gershwin, Hierarchical real-time integrated scheduling of a semiconductor fabrication facility, in *Computer-Aided Manufacturing/Computer-Integrated Manufacturing,* Part 2 of Vol. 61 in the Series *Control and Dynamic Systems,* C.T. Leondes (eds.). Academic Press, New York (1994), 197-241.

[21] H. Yan, S. Lou, S. Sethi, A. Gardel, and P. Deosthali, Testing the robustness of two-boundary control policies in semiconductor manufacturing, to appear in *IEEE Transactions on Semiconductor Manufacturing.*

SURESH P. SETHI
FACULTY OF MANAGEMENT
UNIVERSITY OF TORONTO
TORONTO, CANADA M5S 1V4

Xun Yu Zhou
Department of Systems Engineering and
Engineering Management
The Chinese University of Hong Kong
Shatin, Hong Kong

Studying Flexible Manufacturing Systems via Deterministic Control Problems

Qiji J. Zhu

Abstract

We deduce control equations governed by ordinary differential equations that determine the moments of the inventory levels in a flow control model of a flexible manufacturing system. Then we discuss how to use these control equations to study the expected inventory level set and the optimal inventory level control problem with quadratic cost functions.

1 Introduction

Managing failure prone flexible manufacturing systems is an important and challenging problem. An important aspect of the problem is to manage the inventory levels of various products of such a system. Since the machines in a flexible manufacturing system are subject to random failures the inventory levels are random variables at a given moment. The framework of flow control has been very successful in representing the random inventory levels. We will adopt the basic flow control model described in the recent books by Gershwin [6] and Sethi and Zhang [10]. There are two different routes one can take in studying inventory level management problems: treat the problem as a stochastic control problem or use the distribution functions of the inventory levels to model the problem as a deterministic control problem. There is much discussion on stochastic control models. In contrast deterministic control models are barely used. The major reason seems to be that the deterministic model derived by using the distribution functions of the inventory levels is, in general, a control system involving partial differential equations with controls that are coupled with the partial derivatives of the state variables. Such a control system is very difficult to handle both theoretically and computationally.

The main purpose of this note is to show that for some flexible manufacturing system management problems we can deduce deterministic control models that involve only ordinary differential equations. Since there are abundant results on the theory and computation of control systems governed by ordinary differential equations we will have more tools in our hands to study flexible manufacturing systems.

Our main observation is that the moments of the inventory levels can be determined by a control system governed by ordinary differential equations. Therefore, we can derive a control model involving only ordinary differential equations whenever our major concern is the moments of those random variables. Two important problems that fall in this category are the estimate of the expected inventory set and the optimal control of the inventory levels with a quadratic cost function. In Section 2 we state the general flow control model for a flexible manufacturing system and discuss the relations that determines the first and second moments of the inventory levels. Then we discuss how to apply these relations to the two concrete problems mentioned above in Section 3 and Section 4 respectively. In principle our method also applies to problems that involve higher moments of the inventory levels.

2 Flow Control Model for Flexible Manufacturing Systems and Moment Equations

We consider a general flexible manufacturing system that produces P part types on M machines. We assume that each machine in this system has two states: operational denoted by 1 and breakdown denoted by 0. Denote the state of the mth machine by s_m. Then $(s_1, ..., s_M)$ represents the state of the whole system. Let S be the finite set of system states, i.e., $S = \{(s_1, ..., s_M) : s_m = 0 \text{ or } 1, m = 1, ..., M\}$. Assume that for each state $s \in S$ the system has a corresponding production capacity set of various parts combinations that the system can produce in one time unit denoted by $C(s)$. In this model the M machines are treated as one 'block'. The set $C(s)$ describes the capacity of the whole block and depends only on the state s of the system. For some relatively simple configurations of such an M-machine system, one can determine $C(s)$ according to the capacity of each individual machine. For example, suppose the capacity set of machine m when state is s_m to be $C_m(s_m)$. If the M machines in the manufacturing system are parallel and each part needs only to be processed by one of the machines then $C(s) = \bigcup_{m=1}^{M} C_m(s_m)$. As another example consider a system with M machines in tandem. Suppose each part needs to be processed sequentially by all the machines and assume infinite buffer between machines. Then $C(s) = \bigcap_{m=1}^{M} C_m(s_m)$. We assume that, for all $s \in S$, $C(s)$ are polyhedrons as one usually encounters in practical problems. We assume that the demand rate for the pth part at time t is $d_p(t)$. Then vector $d(t) = (d_1(t), ..., d_P(t))^\top$ represents the demand rate for the whole system at time t. Let $u(t) = (u_1(t), u_2(t), \ldots, u_P(t))^\top$ be the instantaneous production rate of the system at time t and let $x(t) = (x_1(t), x_2(t), \ldots, x_P(t))^\top$ be the inventory levels of the P parts at time t. Denote the state of the system at time t by $s(t)$. We assume that the initial inventory ξ and the initial state α are known and fixed. For such a system an (open loop) control policy will depend on both time and the state of the system. We denote such a control by $u^{s(t)}(t)$. Then the dynamic of the system is

$$\dot{x}(t) = u^{s(t)}(t) - d(t), \quad x(0) = \xi, \quad u^{s(t)}(t) \in C(s(t)). \tag{1}$$

We assume that $\{s(t) : t \geq 0\}$ is a continuous time Markov process with generator $(\lambda_{ab})_{a,b \in S}$ where $\lambda_{aa} := -\sum_{b \in S \setminus \{a\}} \lambda_{ba}$. For this system, given a control policy $u^{s(t)}(t)$, at a fixed moment t, the inventory levels $x(t)$ and the system state $s(t)$ are random variables. These random variables can be characterized by the hybrid probability density functions $f_b(x,t), b \in S$ defined by

$$f_b(x,t)\delta x_1 \delta x_2 \cdots \delta x_P$$
$$= Prob\{x_1 \leq x_1(t) \leq x_1 + \delta x_1, \ldots, x_P \leq x_P(t) \leq x_P + \delta x_P, s(t) = b\}$$
$$+ o(\delta x_1 \cdots \delta x_P). \tag{2}$$

Throughout this paper we make the following technical assumption on $f_b(x,t)$.

(H) For each $b \in S$, the partial derivatives of f_b exist almost everywhere and are (Lebesgue) integrable on R^{P+1}.

Then it is well known that $f_b(x, t)$ satisfies the following partial differential equations (see [4, 5] and also see [6] for an intuitive argument).

$$\frac{\partial f_b(x,t)}{\partial t} + \sum_{p=1}^{P} \frac{\partial f_b(x,t)}{\partial x_p}(u_p^b(t) - d_p(t)) = \sum_{a \in S} \lambda_{ba} f_a(x,t), \quad (3)$$
$$b \in S, \quad u^b(t) \in C(b).$$

Equation (3) is derived under the assumption (H). When the coefficients of equation (3) are "nice", for example, when d_p's are smooth and the controls u_p^b are piecewise constant (as in the widely used hedging point policy), equation (3) determines f_b's (given initial conditions that come from the actual system) that satisfy condition (H). These f_b's will be the density functions. Thus, in essence, assumption (H) requires that the controls involved to be 'nice' functions, e.g., piecewise constant functions. In this sense it would be appropriate to require the control functions to be piecewise constant. However, the control set of measurable functions is much easy to handle. Since any measurable function can be approximated by a piecewise constant function to any given accuracy we will consider measurable control functions in the sequel.

In some flexible manufacturing system management problems we are concerned only with the moments of the inventory levels. For those problems we can integrate (3) to get certain ordinary differential equations that determine the moments of the inventory levels. By doing so we simplify the mathematical model to one that involves only ordinary differential equations. We will need the following notation for conditional moments.

$$m_{qp}(f_b)(t) = \int_{R^P} x_q x_p f_b(x,t) dx, \quad q, p = 1, \ldots, P, \quad (4)$$
$$m_p(f_b)(t) = \int_{R^P} x_p f_b(x,t) dx, \quad p = 1, \ldots, P, \quad (5)$$

and

$$m(f_b)(t) = \int_{R^P} f_b(x,t)dx. \tag{6}$$

We now discuss relations that determine up to the second moments of the inventory levels.

Theorem 2.1 *Assume that, for $b \in S$, $f_b(x,t)$ satisfy condition (H). Then*

$$\frac{dm(f_b)(t)}{dt} = \sum_{a \in S} \lambda_{ba} m(f_a)(t), \tag{7}$$

$$m(f_b)(0) = \begin{cases} 1 & \text{if } b = \alpha, \\ 0 & \text{otherwise.} \end{cases} \tag{8}$$

Proof. Since $C(b)$ is bounded for all $b \in S$, the set $\cup_{b \in S} C(b)$ is bounded. For any given t, $x(t)$ is contained in $t \cup_{b \in S} C(b)$. Therefore, $f_b(x,t) = 0$ when $||x||$ is sufficiently large (more precisely when $x \notin t \cup_{b \in S} C(b)$) for all $b \in S$. Integrating (3) over R^P yields the equation (7) in the lemma. The initial condition comes from the fact that $m(f_b)(0) = Prob\{\alpha = b\}$. ∎

Remark 2.2 *Observe that $m(f_b)(t) = Prob\{s(t) = b\}$. We can conclude that $\sum_{b \in S} m(f_b)(t) = 1$ for all t.*

Theorem 2.3 *Assume that $f_b(x,t), b \in S$ satisfy condition (H). Then*

$$\frac{dm_p(f_b)(t)}{dt} = \sum_{a \in S} \lambda_{ba} m_p(f_a)(t) + m(f_b)(t)(u_p^b(t) - d_p(t)), \tag{9}$$

$$p = 1,\ldots,P, \quad b \in S, \quad u^b(t) \in C(b),$$

$$(m_1(f_b)(0), \ldots, m_P(f_b)(0))^\top = \begin{cases} \xi & \text{if } b = \alpha, \\ 0 & \text{otherwise.} \end{cases} \tag{10}$$

Proof. Integrating by parts we have

$$\int_{R^P} x_p \frac{\partial f_b(x,t)}{\partial x_q} dx = \int_{R^{P-1}} \int_{-\infty}^{\infty} x_p \frac{\partial f_b(x,t)}{\partial x_q} dx_q dx_1 \cdots dx_{q-1} dx_{q+1} \cdots dx_P$$

$$= \begin{cases} -m(f_b)(t) & \text{if } p = q, \\ 0 & \text{if } p \neq q. \end{cases} \tag{11}$$

Multiplying (3) by x_p and integrating with respect to x over R^P we obtain

$$\frac{dm_p(f_b)(t)}{dt} = \sum_{a \in S} \lambda_{ba} m_p(f_a)(t) - \sum_{q=1}^{P} \int_{R^P} x_p \frac{\partial f_b(x,t)}{\partial x_q} dx(u_q^b(t) - d_q(t)),$$

$$p = 1,\ldots,P, \quad b \in S, \quad u^b(t) \in C(b). \tag{12}$$

Then equation (9) is obtained by substituting (11) into (12).

By the definition of $f_b(x,0)$, $f_b(x,0) = 0$ when $b \neq \alpha$ and $f_\alpha(x,0)$ is the density function of the inventory levels at $t = 0$. Therefore, by definition $m_p(f_b)(0) = 0$ when $b \neq \alpha$ and $m_p(f_\alpha)(0)$ is the mean of the inventory level of the pth part. Since the inventory levels at $t = 0$ are fixed and equal to ξ we conclude that $m_p(f_\alpha)(0) = \xi_p$ for $p = 1, ..., P$, and (10) follows. ∎

Theorem 2.4 *Assume that, for $b \in S$, $f_b(x,t)$ satisfy condition (H). Then*

$$\frac{d}{dt}m_{qp}(f_b)(t) = \sum_{a \in S} \lambda_{ba} m_{qp}(f_a)(t) + m_q(f_b)(t)(u_p^b(t) - d_p(t)),$$
$$+ m_p(f_b)(t)(u_q^b(t) - d_q(t)) \qquad (13)$$

$$q, p = 1, \ldots, P, \quad b \in S, \quad u^b(t) \in C(b),$$

$$m_{qp}(f_b)(0) = \begin{cases} \xi_q \xi_p & \text{if } b = \alpha, \\ 0 & \text{otherwise.} \end{cases} \qquad (14)$$

Proof. Again, integrating by parts we have

$$\int_{R^P} x_q x_p \frac{\partial}{\partial x_k} f_b(x,t) dx = \begin{cases} 0 & k \neq q, \ k \neq p, \\ -m_p(f_b)(t) & k = q \neq p, \\ -m_q(f_b)(t) & k = p \neq q, \\ -2m_p(f_b)(t) & k = q = p. \end{cases} \qquad (15)$$

Thus,

$$\frac{d}{dt}m_{qp}(f_b)(t) = \frac{d}{dt}\int_{R^P} x_q x_p f_b(x,t) dx = \int_{R^P} x_q x_p \frac{\partial}{\partial t} f_b(x,t) dx$$
$$= \sum_{a \in S} \int_{R^P} x_q x_p \lambda_{ba} f_a(x,t) dx$$
$$- \sum_{k=1}^{P} \int_{R^P} x_q x_p \frac{\partial}{\partial x_k} f_b(x,t) dx (u_k^b(t) - d_k(t))$$
$$= \sum_{a \in S} \lambda_{ba} m_{qp}(f_a)(t) + m_q(f_b)(t)(u_p^b(t) - d_p(t))$$
$$+ m_p(f_b)(t)(u_q^b(t) - d_q(t)).$$

When $t = 0$ the inventory levels ξ are fixed. We can view the components of ξ as independent "random" variables. Thus, $m_{qp}(f_b)(0) = m_q(f_b)(0) m_p(f_b)(0)$. Combining this decomposition with (10) yields the initial condition. ∎

We can see that, for $b \in S$, functions $m(f_b)$ are determined by Theorem 2.1 provided the initial state α is given. For any practical flexible manufacturing system the initial state of the system is always known. Therefore, we can calculate $m(f_b)$ first for $b \in S$ and treat them as known functions in the rest of the analysis process. For a given control policy $u^{s(t)}(t)$, Theorem 2.3 determines the first moments and Theorem 2.3 and Theorem 2.4 determine the second moments of the inventory levels. Using similar arguments we can also derive equations that determine higher moments of the inventory levels. In the next two sections we discuss how to use Theorems 2.1, 2.3 and 2.4 to study the expected inventory level sets and an optimal inventory level control problem with quadratic cost functions.

3 Estimate the Expected Inventory Level Sets

The expected inventory set of a flexible manufacturing system is the set of all possible expected inventory levels. It is an important measurement of the capacity of the system. When the demand rate $d(t)$ is identically 0 it reduces to the expected production set discussed in [3]. If 0 is not in this expected inventory set then it means that the demand rate exceeds the system capacity. In general the distance of 0 to the boundary of the expected inventory set indicates the room that one has in manipulating the inventory levels. As an extremal case if 0 is on the boundary of the expected inventory set then it means that the demand rate is feasible to the system (in the sense of expectation) but there is no room for any manipulation. The only choice is to operating the machines in the system in its full capacity whenever they are operational according to the demand. In this section we discuss how to estimate the expected inventory set for a flexible manufacturing system. To simplify the exposition, we assume that the initial inventory $\xi = 0$. General result can be derived by a shift of ξ.

We will use the following notation. Let R^P be the P-dimensional Euclidean space. The scalar product of two vectors $x = (x_1, \ldots, x_P)^\top$ and

$y = (y_1, \ldots, y_P)^\top$ in R^P is

$$< x, y > = \sum_{p=1}^{P} x_p y_p. \tag{16}$$

The norm $\|x\|$ of x is its Euclidean norm, i.e., $\|x\| = \sqrt{\sum_{p=1}^{P} x_p^2}$. Let $\mathcal{P}(R^P)$ be the collection of all subsets of R^P. The sum of two subsets E and F of R^P is $E + F = \{x + y \mid x \in E, \ y \in F\}$. The product of a set E in $\mathcal{P}(R^P)$ and a number a is defined by $aE = \{ax : x \in E\}$. The support function $\sigma(x, E)$ of a subset E of R^P is defined, for every $x \in R^P$, by

$$\sigma(x, E) = \sup\{< x, y > : y \in E\}. \tag{17}$$

The convex hull and the closed convex hull of a subset E of R^P are denoted by coE and $\overline{co}E$ respectively. Let $F : [0, t] \to \mathcal{P}(R^P)$ be a multifunction (set-valued function). Following Aumann [2], the integral of F on $[0, t]$ is defined by

$$\int_0^t F(s) ds = \left\{ \int_0^t f(s) ds : f \text{ is integrable and } f(s) \in F(s) \text{ a.e. on } [0, t] \right\}. \tag{18}$$

For any given control policy $u^{s(t)}(t)$ we denote the corresponding overall expected inventory levels by $\mu(t) = (\mu_1(t), \ldots, \mu_P(t))^\top$. Note that, for each p, the moment $m_p(f_b)(t)$ determined by Theorem 2.3 is the mean of the inventory level of part p when $s(t) = b$. We can express μ_p as

$$\mu_p(t) = \sum_{b \in S} m_p(f_b)(t).$$

Taking the sum of (9) over S for all $b \in S$ we obtain

$$\begin{aligned}
\frac{d\mu_p(t)}{dt} &= \sum_{b \in S} \frac{d}{dt} m_p(f_b)(t) \\
&= \sum_{b \in S} \sum_{a \in S} \lambda_{ba} m_p(f_a)(t) + \sum_{b \in S} m(f_b)(t)(u_p^b(t) - d_p(t)) \\
&= \sum_{a \in S} (\sum_{b \in S} \lambda_{ba}) m_p(f_a)(t) + \sum_{b \in S} m(f_b)(t)(u_p^b(t) - d_p(t)) \\
&= \sum_{b \in S} m(f_b)(t)(u_p^b(t) - d_p(t))
\end{aligned} \tag{19}$$

where the last equality is due to $\sum_{b \in S} \lambda_{ba} = 0$. In vector form we have

$$\frac{d\mu(t)}{dt} = \sum_{b \in S} m(f_b)(t)(u^b(t) - d(t)). \tag{20}$$

Integrating (20) from 0 to t and incorporating the initial condition (10) we obtain

$$\begin{aligned} \mu(t) &= \int_0^t \sum_{b \in S} m(f_b)(\tau)(u^b(\tau) - d(\tau)) d\tau \\ &= \int_0^t \sum_{b \in S} m(f_b)(\tau) u^b(\tau) d\tau - \int_0^t d(\tau) d\tau. \end{aligned} \tag{21}$$

where the last equality follows from Remark 2.2. Denote $I(t)$ the expected inventory level set at t. Using the notation of the Aumann's integral and (21) we have

$$I(t) = \sum_{b \in S} \int_0^t m(f_b)(\tau) C(b) d\tau - \int_0^t d(\tau) d\tau. \tag{22}$$

The following theorem further simplifies the expression of $I(t)$.

Theorem 3.1

$$I(t) = \sum_{b \in S} \int_0^t m(f_b)(\tau) d\tau C(b) - \int_0^t d(\tau) d\tau. \tag{23}$$

Theorem 3.1 is an immediate consequence of (22) and the following lemma:

Lemma 3.2 *Let f be an integrable function and let Ω be a compact convex set. Then,*

$$\int_0^t f(\tau) \Omega d\tau = \left(\int_0^t f(\tau) d\tau \right) \Omega. \tag{24}$$

Proof. By Aumann's theorem, the left-hand side of the equality is a convex compact set. It is obvious that so is the right-hand side. To prove the equality we need only to prove that the support function of both sides coincide [9]. This fact follows from the following calculation. For any $x \in R^P$,

$$\sigma(x, \int_0^t f(\tau) \Omega d\tau) = \max\{< x, y > | \, y \in \int_0^t f(\tau) \Omega d\tau\}$$

$$\begin{aligned}
&= \int_0^t f(\tau) \cdot \max\{<x,y> | y \in \Omega\} d\tau \\
&= \int_0^t f(\tau) d\tau \cdot \max\{<x,y> | y \in \Omega\} \\
&= \int_0^t f(\tau) d\tau \cdot \sigma(x, \Omega) \\
&= \sigma(x, \int_0^t f(\tau) d\tau \Omega).
\end{aligned}$$

∎

Now we can use Theorem 3.1 to calculate the expected inventory level set. Observe that as a finite sum of polyhedrons, the expected inventory level set is a polyhedron. A naive approach is to calculate all possible weighted sums of vertices of $C(b)$ as given in the expression of Theorem 3.1. This approach is useful when the numbers of states and vertices of the capacity set corresponding to each state are small. But, many of those weighted sums are not vertices of $I(t)$ and the number of such sums is exponential in the number of the states of the system. Therefore, the following adaptation of the approximation scheme suggested in Zhu, Zhang and He [11], which calculates only vertices of $I(t)$, is more practical in dealing with systems with a large number of states.

Approximation Scheme:

1. Choose a grid of $S^{P-1} \cap R_+^P$ and let L be the set of lattice points, where R_+^P is the positive octave of R^P and S^{P-1} is $(P-1)$-dimensional unit sphere in R^P.

2. For every $l \in L$, determine a $g^b(l) \in C(b)$ such that

$$<l, g^b(l)> = \max_{g \in C(b)} <l, g>. \qquad (25)$$

3. Calculate

$$e(l)(t) = \sum_{b \in S} \int_0^t m(f_b)(\tau) d\tau g^b(l) - \int_0^t d(\tau) d\tau. \qquad (26)$$

Then the convex polyhedron $co\{e(l) \mid l \in L\}$ will approximate the expected inventory set $I(t)$ and $\{e(l)(t) \mid l \in L\}$ is the vertices of this approximation polyhedron.

To implement, we can divide $S^{P-1} \cap R_+^P$ into equal hypercubes under spherical coordinates and let L be the set of the corresponding lattice points as in [11]. The actual construction is as follows:

Choose an integer K according to accuracy requirements and define Q as

$$L = \{l(k_1, \ldots, k_{P-1}) \mid 0 \le k_1, \ldots, k_{P-1} \le K\} \tag{27}$$

where

$$l(k_1, \ldots, k_{P-1}) = \begin{bmatrix} \cos(2\pi k_{P-1}/K) \cdot \cos(2\pi k_{P-2}/K) \ldots \cos(2\pi k_2/K) \cdot \cos(2\pi k_1/K) \\ \sin(2\pi k_{P-1}/K) \cdot \cos(2\pi k_{P-2}/K) \ldots \cos(2\pi k_2/K) \cdot \cos(2\pi k_1/K) \\ \ldots \\ \sin(2\pi k_2/K) \cdot \cos(2\pi k_1/K) \\ \sin(2\pi k_1/K) \end{bmatrix}.$$

By Lemma 4.1 of [11], the relative error caused by such a grid is bounded by $\pi\sqrt{P-1}/K$. For polyhedron $I(t)$, when the grid is fine enough, this algorithm gives accurate vertices.

Control Policy

Comparing with Theorem 2.3, the expression

$$\begin{aligned} e(l)(t) &= \sum_{b \in S} \int_0^t m(f_b)(\tau) d\tau g^b(l) - \int_0^t d(\tau) d\tau \\ &= \sum_{b \in S} \int_0^t m(f_b)(\tau)(g^b(l) - d(\tau)) d\tau. \end{aligned}$$

means that $e(l)(t)$ is the expected inventory levels when using the control $u^{s(t)}(t)$ defined by

$$u^b(t) = g^b(l), \quad b \in S. \tag{28}$$

Therefore, while calculating the expected inventory set, we also get an open loop control policies that can control the system to produce at the boundary of this set.

4 Optimal Inventory Control with Quadratic Cost Function

Suppose that a demand rate is given. When there is enough room to manipulate the flexible manufacturing system it is always desirable to keep the inventory level as "close" to 0 as possible. There are many different criterion for "close" to 0. In this section we discuss an inventory level control problem with quadratic criterion. More specifically, let $x(t) = (x_1(t), ..., x_P(t))^\top$ be the vector representing the inventory levels of the P parts at time t. We consider a quadratic instantaneous cost function $\sum_{p=1}^{P}(a_p x_p^2 + b_p x_p + c_p)$ and seek control policies that minimize the expected cost over a given time interval $[0, T]$. The coefficients a_p will always be positive because $a_p < 0$ means the more backlog or surplus the better which does not make sense.

Using the hybrid distribution functions $f_b(x, t)$ of the inventory levels the expected cost over the time interval $[0, T]$ is

$$\int_0^T \int_{R^P} \sum_{b \in S} \sum_{p=1}^{P} \left(a_p x_p^2(t) + b_p x_p(t) + c_p \right) f_b(x, t) dx dt$$

$$= \int_0^T \sum_{b \in S} \sum_{p=1}^{P} \Big(a_p \int_{R^P} x_p^2(t) f_b(x, t) dx$$
$$+ b_p \int_{R^P} x(t) f_b(x, t) dx + c_p \int_{R^P} f_b(x, t) dx \Big) dt$$

$$= \int_0^T \sum_{b \in S} \sum_{p=1}^{P} \Big(a_p m_{pp}(f_b)(t) + b_p m_p(f_b)(t) + c_p m(f_b)(t) \Big) dt$$

$$= \int_0^T \sum_{p=1}^{P} \left(a_p \sum_{b \in S} m_{pp}(f_b)(t) + b_p \sum_{b \in S} m_p(f_b)(t) + c_p \sum_{b \in S} m(f_b)(t) \right) dt.$$

As observed in Remark 2.2, for all t, $\sum_{b \in S} m(f_b)(t) = 1$. Denote $\nu_p(t) = \sum_{b \in S} m_{pp}(f_b)(t)$. Then the cost function becomes

$$\int_0^T \sum_{p=1}^{P} \left(a_p \nu_p(t) + b_p \sum_{b \in S} m_p(f_b)(t) + c_p \right) dt.$$

We have shown in Section 3 that, for $b \in S$ and $p = 1, ..., P$, $m_p(f_b)$ are determined by equations (9) and initial conditions (10). We now proceed to discuss how to determine ν_p. Set $q = p$ in equations (13) we have, for $b \in S$ and $p = 1, ..., P$,

$$\frac{d}{dt}m_{pp}(f_b)(t) = \sum_{a \in S}\lambda_{ba}m_{pp}(f_a)(t) + 2m_p(f_b)(t)(u_p^b(t) - d_p(t)). \tag{29}$$

Taking the sum of (29) over all $b \in S$ we obtain

$$\begin{aligned}\frac{d\nu_p(t)}{dt} &= \sum_{b \in S}\frac{d}{dt}m_{pp}(f_b)(t) \\ &= \sum_{b \in S}\sum_{a \in S}\lambda_{ba}m_{pp}(f_a)(t) + 2\sum_{b \in S}m_p(f_b)(t)(u_p^b(t) - d_p(t)) \\ &= \sum_{a \in S}(\sum_{b \in S}\lambda_{ba})m_{pp}(f_a)(t) + 2\sum_{b \in S}m_p(f_b)(t)(u_p^b(t) - d_p(t)) \\ &= 2\sum_{b \in S}m_p(f_b)(t)(u_p^b(t) - d_p(t)) \end{aligned} \tag{30}$$

where the last equality is because $\sum_{b \in S}\lambda_{ba} = 0$. The functions $m_p(f_b), p = 1, .., P, b \in S$ in (30) are determined by (9) and (10). Moreover, initial conditions (14) yield $\nu_p(0) = \xi_p^2$, $p = 1, ..., P$. Putting all the information together we are able to formulate the optimal inventory level control problem as the following deterministic optimal control problem:

$(\mathcal{P})$ minimize $\quad\int_0^T \sum_{p=1}^P (a_p\nu_p(t) + b_p\sum_{b \in S}m_p(f_b)(t) + c_p)dt$

subject to $\quad\dfrac{d\nu_p(t)}{dt} = 2\sum_{b \in S}m_p(f_b)(t)(u_p^b(t) - d_p(t)),$

$\nu_p(0) = \xi_p^2, p = 1, ..., P,$

$\dfrac{dm_p(f_b)(t)}{dt} = \sum_{a \in S}\lambda_{ba}m_p(f_a)(t) + m(f_b)(t)(u_p^b(t) - d_p(t)),$

$p = 1, \ldots, P, \quad b \in S,$

$(m_1(f_b)(0), ..., m_P(f_b)(0))^\top = \begin{cases} \xi & \text{if } b = \alpha, \\ 0 & \text{otherwise,} \end{cases}$

$u^b(t) = (u_1^b(t), ..., u_P^b(t))^\top \in C(b).$

Problem $\mathcal{P}$ is a classical optimal control problem. It is linear if and only if, for all $p = 1, ..., P$, $a_p = 0$. However, that corresponding to the linear instantaneous cost functions on the inventory levels which are never useful in practical flexible manufacturing systems (see [6, 10] for discussions on the cost functions). Therefore, we are essentially facing a nonlinear control problem.

For such a control problem explicit solutions are not to be expected. However, many numerical methods are available (cf. [2, 7]). Moreover, the control equations in problem $\mathcal{P}$ is bilinear. There is much study on the qualitative aspects of such bilinear control systems since the 1970's. We refer to [8] and the references therein for those discussions.

We conclude this section with a discussion of the quadratic cost function. To simplify the notation we consider the instantaneous cost corresponding to part p and omit the subscript p, i.e., $ax^2 + bx + c, a > 0$. The simplest useful form of such quadratic cost function is of course ax^2. This form impose equal (quadratic increasing) instantaneous cost on both surplus and backlog. In practice the cost caused by backlog (penalty) is usually higher than that of surplus. Therefore the most common cost function adopted in flexible manufacturing management problems are of the form

$$c^+ \max(0, x) + c^- \max(0, -x) \tag{31}$$

where $c^- > c^+ > 0$ are constants. The method discussed in this paper is not directly applicable to problems with cost function (31). However, for an actual production system there are limits on the maximum backlog, say $l < 0$, and maximum surplus, say $L > 0$. Therefore the cost (31) can be considered only defined on the interval $[l, L]$. On this interval $[l, L]$ we can approximate (31) by a quadratic function so that the problem discussed in this section represents an approximation to the inventory control problem with cost function (31). Greater accuracy can be achieved by using higher order polynomials to approximate the cost (31). Certainly this will yield problems that involving higher order moments of the inventory levels. The approach discussed in this paper will still apply.

5 Conclusion

We observe that the moments of the inventory levels of a flexible manufacturing system can be determined by control systems governed by ordinary differential equations. For problems of estimate the expected inventory sets and optimal inventory level control with quadratic cost function this method yields linear and bilinear control problems. These control models for the flexible manufacturing system management problems enable us to use many tools available in the control system theory for both qualitative and numerical analysis. In principle, our approach also applies to problems that involves higher moments of the inventory levels. It is an interesting topic for future research.

References

[1] R. J. Aumann, *Integrals of Set-Valued Function*, J. Math. Anal. Appl. Vol. 12, pp. 1-12, 1965.

[2] R. Bellman, *Introduction to the Mathematical Theory of Control Processes*, Vol. II. Nonlinear Processes, Academic Press, New York, 1971.

[3] E. K. Boukas and Q. J. Zhu, *Production set of failure prone flexible manufacturing systems*, Research report, GERAD, Montreal, 1993.

[4] M. H. A. Davis, *Markov Models and Optimization*, Chapman and Hall, New York, 1993.

[5] J. L. Doob, *Stochastic Process*, John Wiley & Sons, Inc., New York, 1953.

[6] S. B. Gershwin, *Manufacturing Systems Engineering*, Prentice Hall, 1993.

[7] W. A. Gruver and E. Sachs, *Algorithmic methods in optimal control*, Research notes in math. Vol. 47, Pitman, Boston, 1980.

[8] D. Q. Mayne and R. W. Brockett, *Geometric Methods in System Theory*, Proceedings of the NATO Advanced Study Institute, London, 1973, D. Reidel Publishing, Boston, 1973.

[9] R. T. Rockafellar, *Convex Analysis*, Princeton Mathematics Ser. Vol. 28, Princeton University Press, 1970.

[10] S.P. Sethi and Q. Zhang, *Hierarchical Decision Making in Stochastic Manufacturing Systems*, Birkhauser, Boston, 1994.

[11] Q. J. Zhu, N. Zhang and Y. He, *Algorithm for Determining the Reachability Set of a Linear Control System*, Journal of Optimization Theory and Applications, Vol. 72, pp. 333-353, 1992.

Acknowledgement: I thank the referee for his/her constructive comments on an early version of the paper.

QIJI J. ZHU
DEPARTMENT OF MATHEMATICS AND STATISTICS
WESTERN MICHIGAN UNIVERSITY
KALAMAZOO, MI 49008

On Robust Design for a Class of Failure Prone Manufacturing Systems

E.K. Boukas,[*] Q. Zhang[†] and G. Yin[‡]

Abstract

This paper concerns itself with the development of robust production and maintenance planning in manufacturing systems. Taking robustness into consideration, a new model for failure prone flexible manufacturing system with machines subject to breakdown and repair is proposed and analyzed. The machine capacity is assumed to be a Markov chain with finite state space, whereas the demand rate is treated as a bounded disturbance that can be deterministic or stochastic as long as it is adapted with respect to the Markov chain representing the machine capacity. The control variables are the rate of maintenance and the rate of production, and the cost function has a minimax structure. Under suitable conditions, it is shown that the value function is locally Lipschitz and satisfies a Hamilton-Jacobi-Isaacs (HJI) equation. A sufficient condition for optimal control is obtained. Finally, an algorithm is given for solving the optimal control problem numerically.

Key words: flexible manufacturing system, controlled Markov process, robust production and maintenance planning, Hamilton-Jacobi-Isaacs equation, viscosity solution.

1 Introduction

We develop a new model for a class of failure prone flexible manufacturing systems. Continuing along the line of our previous investigation, the distinct features of the model include: the addition of the maintenance activities, the

---

[*]Research of this author was supported by the Natural Sciences and Engineering Research Council of Canada under grants OGP0036444 and FCAR NC0271 F.

[†]Research of this author was supported in part by the Natural Sciences and Engineering Research Council of Canada under grant A4169 and in part by the Office of Naval Research under Grant N00014-96-1-0263.

[‡]Research of this author was supported in part by the National Science Foundation under grant DMS-9224372, and in part by Wayne State University.

assumed bounded disturbance as the rate of demand and the minimax cost structure. The emphasis is on the robust production and maintenance planning and design.

Realizing the needs and facing the challenges of the today's competitive market, the study of flexible manufacturing systems has witnessed rapid progress in recent years. Most manufacturing systems are large and complex, and require sophisticated modern technology to control their operations. The underlying machines are usually prone to failures, which depend mainly on the usage. To formulate such dependence one often considers the machine capacity as a Markov chain with finite state space, governed by a generator that depends on the production rates; see Hu et al [10], Soner [13], and Sethi and Zhang [12]. In practice, preventive maintenance is used to reduce the rate of machine failures. The modeling of preventive maintenance and related topics was initiated by Boukas and continued by Boukas et al. (see [2, 3, 4, 5]).

By preventive maintenance, we mean the activities such as lubrication, routine adjustments etc. Such an effort clearly reduces the aging rate of the machines and improves the productivity of the systems. However, there will be cost associated with such procedures. Since the maintenance will interrupt the production process, we model the interruption as a part of the control constraints. Our objective is to find the rate of maintenance and the rate of production that minimize the overall costs of inventory/shortage, production, and maintenance. We formulate the optimal production and maintenance scheduling problem as a controlled piecewise deterministic Markov process optimization problem.

A reality check reveals that the demand rate is usually not known and often appeared in form as a bounded disturbance process. It may be deterministic or stochastic. Frequently, one asks the question: What should be the first priority, optimality or stability? Since most manufacturing systems are large complex systems, it is very difficult to establish accurate mathematical models to describe these systems, and modeling error is inevitable. A natural consideration then is to stress the robustness issue. In practice, an optimal policy for a subdivision of a large corporation is usually not an optimal policy for the whole corporation. Therefore, optimal solutions with the usual discounted cost criteria may not be desirable or meaning less. An alternative approach is to consider robust controls that emphasize system stability rather than optimality. Robust control design is particularly important in these manufacturing systems with unfavorable disturbances including: (1) unfavorable internal disturbances – usually associated with unfavorable machine failures; (2) unfavorable external disturbances – usually associated with unfavorable fluctuations in demand. In this paper, we focus our attention to the second type of disturbances. Our main objective is to consider controls that are robust enough to attenuate uncertain disturbances including modeling errors, and to achieve the system stability.

In this paper we formulate the robust production and maintenance scheduling problem as a minimax stochastic control problem. The machine capacity is modelled as a finite state Markov chain with a generator that depends on the aging rate, production rate, and maintenance rate. The demand rate is modelled as an unknown disturbance process. Consider the objective function that is the supremum of the expected discounted cost over all demand processes. We aim to find control policies that minimize the worst case scenario. The advantage of considering such a cost functional is that we do not need the information regarding the probability distribution of the demand facing the system. Such an approach is referred to as a minimax approach, and is closely related to the H^∞-control of a dynamic system; see Başar and Bernhard [1] for more details. The idea behind this approach is to design a control that is robust enough to attenuate the disturbances facing the system and to deal with the worst situation that might arise. By and large, the optimal solution for the robust control problem will provide the most conservative optimal control of the problem.

Using the dynamic programming approach, it is shown that the value function satisfies the associated Hamilton-Jacobi-Isaacs (HJI) equation in the sense of viscosity solutions. The HJI equation has a differential game interpretation. We prove the Lipschitz property of the value function, which guarantees the value function is almost everywhere differentiable, obtain a verification theorem that provides a sufficient condition for optimal control policies, and provide an approximation algorithm to solve the optimal control problem numerically.

This remainder of the paper is arranged as follows. In the next section, we formulate the problem of the robust production and preventive maintenance planning. In Section 3, we show that the value function is locally Lipschitz and is the unique viscosity solution to the HJI equation. Then based on the uniqueness of the solution, a verification theorem is derived. In Section 4, we give an algorithm for solving the optimal policy numerically. Finally some concluding remarks are issued.

2 Problem formulation

Consider a manufacturing system that consists of M machines and produces P different part types. Assume that the machines are subject to random failures and repairs. The machines can be either operational, or under repair. When a machine is operational, it can produce any part type and when it is under repair nothing is produced. Suppose that the capacity of the machines is a finite state Markov chain.

Let $\alpha_i(t)$ denote the capacity of machine i at time t and $\{0, 1, \ldots, \mu_i\}$ the state space of the process $\alpha_i(t)$. Then, the vector process $\alpha(t) = (\alpha_1(t), \ldots, \alpha_M(t)) \in \mathcal{M}$ and

$$\mathcal{M} = \{0, 1, \ldots, \mu_1\} \times \{0, 1, \ldots, \mu_2\} \times \cdots \times \{0, 1, \ldots, \mu_M\}$$

describes the state process of the global manufacturing system. The set $\mathcal{M}$ has $p = (\mu_1 + 1) \cdots (\mu_M + 1)$ states and is denoted by $\mathcal{M} = \{\alpha^1, \ldots, \alpha^p\}$.

Let $u(t) \in \mathbb{R}^P$ and $w(t) \in \mathbb{R}^M$ denote the vectors of the production rate and the maintenance rate of the system at time t, respectively. We use $a_i(t)$ to represent the age of the ith machine at time t and $a(t) = (a_1(t), \ldots, a_M(t))$ the vector age of the system. We use $x_j(t)$ to represent the inventory/shortage level of part type j at time t and $x(t) = (x_1(t), \ldots, x_P(t))$ the vector of inventory/shortage level of the system. Let $z(t) \in \mathbb{R}^P$ denote the vector representing the demand rate. The inventory/shortage levels and machine ages of the system are described by the following differential equations:

$$\dot{x}(t) = u(t) - z(t), \quad x(0) = x$$
$$\dot{a}(t) = m(\alpha(t), u(t), w(t)), \quad a(0) = a \tag{1}$$

where $u(t) = \sum_{i=1}^{M} u_i(t)$ and $u_i(t) = (u_{i1}, \ldots, u_{iP})'$ (here A' denotes the transpose of a vector A) is the instantaneous vector production rate of the ith machine (u_{ij} represents the production rate of part type j of machine i). The function $m(\cdot) = (m_1(\cdot), \ldots, m_M(\cdot))'$ in (1) represents the aging rate of the machines as a function of $\alpha(t)$, $u(t)$, and $w(t)$. We assume that for each i, $m_i(\cdot) \geq 0$, $m_i(\alpha, u, w) = 0$ when $\alpha_i = 0$ (i.e., the ith machine is under repair), and $m(\cdot)$ is linear in (u, w). More specifically, we consider

$$m(\alpha, u, w) = K_1(\alpha)u - K_2(\alpha)w$$

with matrices $K_1(\alpha)$ and $K_2(\alpha)$ of appropriate dimensions. The ith rows of $K_1(\alpha)$ and $K_2(\alpha)$ are 0 whenever $\alpha_i = 0$.

We consider the demand rate $z(\cdot)$ to be an unknown disturbance process fluctuating in a compact set of an Euclidean space. Let $\mathcal{Z}$ denote the set of all such demand rates.

The control variables under consideration are the production rate u and the maintenance rate w. The maintenance interrupts the production process and uses a fraction of machine up times. As discussed in Boukas et al. [6], if the machine state is $\alpha = (\alpha_1, \ldots, \alpha_M)$ then the control constraints can be formulated as:

$$b_{i1}u_{i1} + \cdots + b_{iP}u_{iP} + h_i w_i \leq \alpha_i, \quad i = 1, \ldots, M$$

for nonnegative constants b_{ij} and h_i, $i = 1, 2, \ldots, M$, $j = 1, 2, \ldots, P$. Moreover, $u_{ij}(t) \geq 0$ and $w_i(t) \geq 0$. Namely, no negative production or maintenance will be allowed. Noticing the aging rate is always nonnegative, in addition to the above constraints, we also assume that for each $\alpha \in \mathcal{M}$, for each u and each w,

$$K_1(\alpha)u - K_2(\alpha)w \geq 0.$$

Let $\alpha(t) \in \mathcal{M}$ be a finite state Markov process. Let $\Gamma(\alpha(t))$ denote the following control set:

$$\Gamma(\alpha) = \Gamma_0(\alpha) \bigcap \{K_1(\alpha)u - K_2(\alpha)w \geq 0\},$$

where

$$\Gamma_0(\alpha) = \left\{ (u, w) : u = \sum_{i=1}^{M} u_i, u_i = (u_{i1}, \ldots, u_{iP})', w = (w_1, \ldots, w_M)', \right.$$
$$\left. \sum_{j=1}^{P} b_{ij} u_{ij} + t_i w_i \leq \alpha_i, u_{ij} \geq 0, w_i \geq 0, i = 1, 2, \ldots, M \right\}.$$

To proceed, we first define admissible controls and admissible feedback controls, respectively.

Definition 1. A control $(u(\cdot), w(\cdot)) = \{(u(t), w(t)) : t \geq 0\}$ is *admissible* if:

(i) $(u(\cdot), w(\cdot))$ is adapted to the σ-algebra generated by $\alpha(\cdot)$, denoted as $\sigma\{\alpha(s) : 0 \leq s \leq t\}$, and

(ii) $(u(t), w(t)) \in \Gamma(\alpha(t))$ for all $t \geq 0$.

Let $\mathcal{A}$ denote the set of all admissible controls.

Definition 2. A function $(u(x, a, \alpha), w(x, a, \alpha))$ is an admissible *feedback* control, if

(i) for any given initial x, the following system of equations has a unique solution $(x(\cdot), a(\cdot))$:

$$\dot{x}(t) = u(x(t), a(t), \alpha(t)) - z(t), \quad x(0) = x$$

$$\dot{a}(t) = m(\alpha(t), u(t), w(t)), \quad a(0) = a$$

and

(ii) $(u(\cdot), w(\cdot)) = (u(x(\cdot), a(\cdot), \alpha(\cdot)), w(x(\cdot), a(\cdot), \alpha(\cdot))) \in \mathcal{A}$.

For any given $(u(\cdot), w(\cdot)) \in \mathcal{A}$ and $z(\cdot) \in \mathcal{Z}$, let

$$J(x, a, \alpha, u(\cdot), w(\cdot), z(\cdot)) = E \int_0^\infty e^{-\rho t} G(x(t), a(t), u(t), w(t)) dt,$$

where $\rho > 0$ is a constant representing the discount rate and x, a, and α are the initial values of $x(t)$, $a(t)$, and $\alpha(t)$, respectively.

The objective of the problem is to find a control policy $(u(\cdot), w(\cdot)) \in \mathcal{A}$ that minimizes the following cost functional:

$$\hat{J}(x, a, \alpha, u(\cdot), w(\cdot)) := \sup_{z(\cdot) \in \mathcal{Z}} J(x, a, \alpha, u(\cdot), w(\cdot), z(\cdot)).$$

Define the value function as the minimum of the cost over $(u(\cdot), w(\cdot)) \in \mathcal{A}$, i.e.,

$$v(x, a, \alpha) = \inf_{(u(\cdot), w(\cdot)) \in \mathcal{A}} \hat{J}(x, a, \alpha, u(\cdot), w(\cdot)).$$

We make the following assumptions throughout this paper.

(A1) $G(x, a, u, w)$ is continuous and there exist nonnegative constants C_g and k_g such that for all x, a, u, and w,

$$0 \leq G(x, a, u, w) \leq C_g(1 + |x|^{k_g} + |a|^{k_g})$$

and all x, a, u, w, x', and a',

$$|G(x, a, u, w) - G(x', a', u, w)|$$

$$\leq C_g(1 + |x|^{k_g} + |x'|^{k_g} + |a|^{k_g} + |a'|^{k_g})(|x - x'| + |a - a'|).$$

(A2) $\alpha(t) \in \mathcal{M}$ is a Markov chain generated by $Q(a, u, w) = (q_{\alpha\beta}(a, u, w))$ such that $q_{\alpha\beta}(a, u, w) \geq 0$ for $\alpha \neq \beta$ and

$$q_{\alpha\alpha}(a, u, w) = -\sum_{\beta \neq \alpha} q_{\alpha\beta}(a, u, w).$$

Moreover, $q_{\alpha\beta}(a, u, w)$ is bounded, Lipschitz in a, and $|q_{\alpha\alpha}(a, u, w)| \geq c_0 > 0$, for some constant $c_0 > 0$.

(A3) The demand $z(\cdot)$ is $\sigma\{\alpha(s) : 0 \leq s \leq t\}$ adapted. For all $t \geq 0$, $z(t) \geq 0$ and $z(t) \in \Gamma_z$, a compact subset of $\mathbb{R}^P$.

Remark: Assumption (A1) indicates that the running cost function satisfies certain growth condition in (x, a). Moreover, it is locally Lipschitz in (x, a).

Assumption (A2) is a condition on the Markov chain involved. Among other things, the diagonal elements of the Q matrix are strictly negative. Note that the generator $Q(a, u, w)$ depends on the control variables as well as state variables of the system under consideration.

Finally, (A3) tells us that the demand is non-anticipative and adapted to the machine capacity process. This condition is needed in the analysis of the associated dynamic programming equations. The interpretation of this condition is that the demand may behave in response to the fluctuation of the machine states. It should be noted that the model we have formulated in the paper does not require a specified probabilistic structure of $z(\cdot)$. Thus, our formulation covers a wide spectrum of demand processes, including both deterministic and random processes as long as they are bounded and not depending on future.

The Markov chain $\alpha(\cdot)$ can be constructed via the piecewise-deterministic approach introduced by Davis [8]. Let us summarize the construction of the Markov process below.

Let $0 = \tau_0 < \tau_1 < \cdots < \tau_k < \cdots$ denote a sequence of jump-times of $\alpha(t)$. Let $\alpha(0) = \alpha \in \mathcal{M}$. Then, on the interval $[\tau_0, \tau_1)$, we have

$$\dot{x}(t) = u(t) - z(t), \quad x(0) = x,$$

$$\dot{a}(t) = m(\alpha, u(t), w(t)), \quad a(0) = a,$$

for deterministic $(u(t), w(t)) \in \Gamma(\alpha)$, $t \geq 0$. The first jump-time τ_1 has the probability distribution

$$P(\tau_1 \geq t) = \exp\left\{\int_0^t q_{\alpha\alpha}(a(s), u(s), w(s))ds\right\}.$$

Then the post-jump location of $\alpha(t) = \alpha^j \in \mathcal{M}$ $(\alpha^j \neq \alpha)$ is given by

$$P(\alpha(\tau_1) = \alpha^j | \tau_1) = \frac{q_{\alpha\alpha^j}(a(\tau_1), u(\tau_1), w(\tau_1))}{-q_{\alpha\alpha}(a(\tau_1), u(\tau_1), w(\tau_1))}.$$

In general, on the interval $[\tau_k, \tau_{k+1})$, we have

$$\dot{x}(t) = u(t) - z(t), \ x(\tau_k) = x(\tau_k^-),$$
$$\dot{a}(t) = m(\alpha(\tau_k), u(t), w(t)), \ a(\tau_k) = a(\tau_k^-)$$

for deterministic $(u(t), w(t)) \in \Gamma(\alpha(\tau_k))$, $t \geq \tau_k$. The jump-time τ_{k+1} has the conditional probability distribution

$$P(\tau_{k+1} - \tau_k \geq t | \tau_1, \ldots, \tau_k, \alpha(\tau_1), \ldots, \alpha(\tau_k))$$
$$= \exp \left\{ \int_{\tau_k}^{t+\tau_k} q_{\alpha(\tau_k)\alpha(\tau_k)}(a(s), u(s), w(s)) ds \right\}.$$

The post-jump location of $\alpha(t) = \alpha^j \in \mathcal{M}$ $(\alpha^j \neq \alpha(\tau_k))$ is given by

$$P(\alpha(\tau_{k+1}) = \alpha^j | \tau_1, \ldots, \tau_k, \tau_{k+1}, \alpha(\tau_1), \ldots, \alpha(\tau_k))$$
$$= \frac{q_{\alpha(\tau_k)\alpha^j}(a(\tau_{k+1}), u(\tau_{k+1}), w(\tau_{k+1}))}{-q_{\alpha(\tau_k)\alpha(\tau_k)}(a(\tau_{k+1}), u(\tau_{k+1}), w(\tau_{k+1}))}.$$

Then, $\alpha(t)$ is a Markov process. Moreover, the process

$$f(\alpha(t)) - f(\alpha(0)) - \int_0^t Q(a(r), u(r), w(r)) f(\alpha(r)) dr$$

is a martingale for any function f on $\mathcal{M}$.

Let H be a real-valued functional defined by:

$$H(x, a, \alpha, v(x, a, \cdot), r_1, r_2, u, w, z)$$
$$= (u - z)r_1 + m(\alpha, u, w)r_2 + G(x, a, u, w)$$
$$+ \sum_{\beta \neq \alpha} q_{\alpha\beta}(a, u, w)[v(x, a, \beta) - v(x, a, \alpha)].$$

Then, the HJI equation associated with the problem can be formally written as follows:

$$\rho v(x, a, \alpha) = \min_{(u,w) \in \Gamma(\alpha)} \max_{z \in \Gamma_z} H(x, a, \alpha, v(x, a, \cdot), v_x(x, a, \alpha), v_a(x, a, \alpha), u, w, z). \tag{2}$$

Remark: This HJI equation has a differential game interpretation. Actually, it associates with a two player, zero-sum differential game in which the system controller plays the role of one player and the nature (demand rate) plays the role of the other player. The controller tries to minimize the objective function $J(x, a, \alpha, u(\cdot), w(\cdot), z(\cdot))$ while, the nature tries to maximize it. Therefore the problem in this paper can also be regarded as a differential game problem in which the controller plays with the nature. The control designed is to act against the worst case of the nature.

It should be noticed that the problem considered in the paper is different from a differential game. A differential game usually requires flows between the two players. In our model, we only consider one way flow because the player representing the market demand is out of the control of a plant manager. Moreover, since the random process $\alpha(\cdot)$ is a finite state Markov chain, the model and the associated HJI equations make sense without using the Elliott-Kalton formulation of differential games.

Note that the HJI equation (2) involves the first order partial derivatives of the value function under no guarantee that those partial derivatives would exist. In fact, one can always construct an example in which the value function does not have partial derivatives at certain points. In order to deal with such nondifferentiability of the value function, the notion of viscosity solutions– introduced by Crandall and Lions is usually used. In what follows, we briefly describe the concept. For more information and discussion on viscosity solutions, the readers are referred to the recent book by Fleming and Soner [9].

For each $r_1, h_1 \in \mathbb{R}^P$ and $r_2, h_2 \in \mathbb{R}^M$, let

$$\Phi(x, a, \alpha, r_1, r_2, h_1, h_2) = \frac{v(x + h_1, a + h_2, \alpha) - v(x, a, \alpha) - h_1 r_1 - h_2 r_2}{|h_1| + |h_2|}.$$

We define the superdifferential $D^+v(x, a, \alpha)$ and subdifferential $D^-v(x, a, \alpha)$

as follows:

$$D^+v(x,a,\alpha) = \{r = (r_1, r_2) \in \mathbb{R}^P \times \mathbb{R}^M : \limsup_{(h_1,h_2)\to 0} \Phi(x,a,\alpha,r_1,r_2,h_1,h_2) \leq 0\},$$

$$D^-v(x,a,\alpha) = \{r = (r_1, r_2) \in \mathbb{R}^P \times \mathbb{R}^M : \liminf_{(h_1,h_2)\to 0} \Phi(x,a,\alpha,r_1,r_2,h_1,h_2) \geq 0\}.$$

Definition 3. We say that $v(x,a,\alpha)$ is a *viscosity solution* to (2) if:

(a) $v(x,a,\alpha)$ is continuous and $|v(x,a,\alpha)| \leq C(1 + |x|^{k_g} + |a|^{k_g})$ for some constant C.

(b) For all $r = (r_1, r_2) \in D^+v(x,a,\alpha)$,

$$\rho v(x,a,\alpha) \leq \min_{(u,w)\in\Gamma(\alpha)} \max_{z\in\Gamma_z} H(x,a,\alpha,v(x,a,\cdot),r_1,r_2,u,w,z).$$

(c) For all $r = (r_1, r_2) \in D^-v(x,a,\alpha)$,

$$\rho v(x,a,\alpha) \geq \min_{(u,w)\in\Gamma(\alpha)} \max_{z\in\Gamma_z} H(x,a,\alpha,v(x,a,\cdot),r_1,r_2,u,w,z).$$

It is easy to see that if $v(x,a,\alpha)$ is differentiable at (x_0, a_0), then $D^+v(x_0,a_0,\alpha) = D^-v(x_0,a_0,\alpha) = \{v_x(x_0,a_0,\alpha), v_a(x_0,a_0,\alpha)\}$. Hence, the equality in the HJI equation (2) holds at points whenever the value function $v(x,a,\alpha)$ is differentiable.

In the next section, we show that the value function is indeed a viscosity solution to the HJI equation.

3 Main results

In this section, we first show that the value function $v(x,a,\alpha)$ is locally Lipschitz and is the only viscosity solution to the HJI equation. We then prove a verification theorem that provides a sufficient condition for optimal control.

Theorem 1 (HJI equation). (i) $0 \leq v(x,a,\alpha) \leq C(1 + |x|^{k_g} + |a|^{k_g})$.

(ii) $|v(x,a,\alpha) - v(x',a',\alpha)| \leq C(1 + |x|^{k_g} + |x'|^{k_g} + |a|^{k_g} + |a'|^{k_g})(|x - x'| + |a - a'|)$.

(iii) $v(x, a, \alpha)$ *is the only viscosity solution to the HJI equation* (2).

Remark: Note that the Lipschitz property in (ii) implies the value function is differentiable almost everywhere (see [12]). Consequently, the value function $v(x, a, \alpha)$ satisfies the HJI equation (2) almost everywhere in the classical sense. Namely, the value function is a generalized solution to the HJI equation. However, the HJI equation might have more than one generalized solutions. The advantage of a viscosity solution is that it characterizes uniquely the solution to the HJI equation.

Proof of Theorem 1. By Assumption (A1), $G(x, a, u, w)$ is nonnegative and has the growth rate at most equal to $(1 + |x|^{k_g} + |a|^{k_g})$ since the controls u and w are all bounded variables. It follows that the value function is also nonnegative and has the growth rate less than a multiple of $(1 + |x|^{k_g} + |a|^{k_g})$. This proves (i).

We now prove (ii). Let

$$v^0(x, a, \alpha) = \inf_{(u(\cdot), w(\cdot)) \in \mathcal{A}} \sup_{z(\cdot) \in \mathcal{Z}} J^0(x, a, \alpha, u(\cdot), w(\cdot), z(\cdot)),$$

where $J^0(x, a, \alpha, u(\cdot), w(\cdot), z(\cdot)) = J(x, a, \alpha, u(\cdot), w(\cdot), z(\cdot))$ when there is no jumps in $\alpha(t)$, i.e., $\alpha(t) = \alpha$ for all $t \geq 0$.

Next we define J^N and v^N inductively. For any $N = 1, 2, \ldots$, $(u(\cdot), w(\cdot)) \in \mathcal{A}$, and $z(\cdot) \in \mathcal{Z}$, let τ denote the first jump time of $\alpha(t)$,

$$J^N(x, a, \alpha, u(\cdot), w(\cdot), z(\cdot)) = E\left(\int_0^\tau e^{-\rho t} G(x(t), a(t), u(t), w(t)) dt \right.$$
$$\left. + e^{-\rho \tau} v^{N-1}(x(\tau), a(\tau), \alpha(\tau)) \right)$$

which can be rewritten after using the fact τ is exponentially disturbed and $P(\tau < \infty) = 1$, integrating by part and using the assumption (A2) as follows:

$$J^N(x, a, \alpha, u(\cdot), w(\cdot), z(\cdot)) = \int_0^\infty e^{-\rho t} \left(e^{-\int_0^t \lambda(r) dr} \right)$$
$$\cdot \left(G(x(t), a(t), u(t), w(t)) + \lambda(t) \sum_{\beta \neq \alpha} v^{N-1}(x(t), a(t), \beta) p_\beta(t) \right) dt,$$

where

$$\lambda(t) = -q_{\alpha\alpha}(a(t), u(t), w(t)), \text{ and } p_\beta(t) = \frac{q_{\alpha\beta}(a(t), u(t), w(t))}{-q_{\alpha\alpha}(a(t), u(t), w(t))}.$$

Define

$$v^N(x, a, \alpha) = \inf_{(u(\cdot), w(\cdot)) \in \mathcal{A}} \sup_{z(\cdot) \in \mathcal{Z}} J^N(x, a, \alpha, u(\cdot), w(\cdot), z(\cdot)).$$

We will show in the following that:

(a) Each $v^N(x, a, \alpha)$ is locally Lipschitz in (x, a);

(b) $v(x, a, \alpha) = \lim_{N \to \infty} v^N(x, a, \alpha)$;

(c) $v(x, a, \alpha)$ is also locally Lipschitz in (x, a).

We verify (a) first. For any $x_1, x_2, a_1,$ and a_2, let $x_1(t), x_2(t), a_1(t)$ and $a_2(t)$ denote the system trajectories under $u(\cdot), w(\cdot),$ and $z(\cdot)$ with $x_1(0) = x_1$, $x_2(0) = x_2$, $a_1(0) = a_1$, and $a_2(0) = a_2$. Then, in view of Assumption (A1), we have

$$|J^0(x_1, a_1, \alpha, u(\cdot), w(\cdot), z(\cdot)) - J^0(x_2, a_2, \alpha, u(\cdot), w(\cdot), z(\cdot))|$$

$$\leq E \int_0^\infty e^{-\rho t} |G(x_1(t), a_1(t), u(t), w(t)) - G(x_2(t), a_2(t), u(t), w(t))| dt$$

$$\leq E \int_0^\infty e^{-\rho t} (1 + |x_1(t)|^{k_g} + |x_2(t)|^{k_g} + |a_1(t)|^{k_g} + |a_2(t)|^{k_g})$$

$$\times (|x_1(t) - x_2(t)| + |a_1(t) - a_2(t)|) dt.$$

Note that

$$|x_i(t)| \leq C(1 + t + |x_i|) \text{ and } |a_i(t)| \leq C(1 + t + |a_i|)$$

for $i = 1, 2$ and

$$\begin{aligned} x_1(t) &= x_1 + \int_0^t (u(s) - z(s)) ds, \quad a_1(t) = a_1 + \int_0^t m(\alpha, u(s), w(s)) ds \\ x_2(t) &= x_2 + \int_0^t (u(s) - z(s)) ds, \quad a_2(t) = a_2 + \int_0^t m(\alpha, u(s), w(s)) ds. \end{aligned} \quad (3)$$

These imply that

$$|x_1(t) - x_2(t)| = |x_1 - x_2| \text{ and } |a_1(t) - a_2(t)| = |a_1 - a_2|.$$

Therefore,

$$|J^0(x_1, a_1, \alpha, u(\cdot), w(\cdot), z(\cdot)) - J^0(x_2, a_2, \alpha, u(\cdot), w(\cdot), z(\cdot))|$$

$$\leq E \int_0^\infty e^{-\rho t} C(1 + t^{k_g} + |x_1|^{k_g} + |x_2|^{k_g} + |a_1|^{k_g} + |a_2|^{k_g})$$

$$\times (|x_1 - x_2| + |a_1 - a_2|) dt$$

$$\leq C_0 (1 + |x_1|^{k_g} + |x_2|^{k_g} + |a_1|^{k_g} + |a_2|^{k_g})(|x_1 - x_2| + |a_1 - a_2|).$$

Hence,

$$|v^0(x_1, a_1, \alpha) - v^0(x_2, a_2, \alpha)|$$

$$\leq \sup_{(u(\cdot), w(\cdot)) \in \mathcal{A}} \left| \sup_{z(\cdot) \in \mathcal{Z}} J^0(x_1, a_1, \alpha, u(\cdot), w(\cdot), z(\cdot)) \right.$$

$$\left. - \sup_{z(\cdot) \in \mathcal{Z}} J^0(x_2, a_2, \alpha, u(\cdot), w(\cdot), z(\cdot)) \right|$$

$$\leq \sup_{(u(\cdot), w(\cdot)) \in \mathcal{A},\ z(\cdot) \in \mathcal{Z}} \left| J^0(x_1, a_1, \alpha, u(\cdot), w(\cdot), z(\cdot)) \right.$$

$$\left. - J^0(x_2, a_2, \alpha, u(\cdot), w(\cdot), z(\cdot)) \right|$$

$$\leq C_0 (1 + |x_1|^{k_g} + |x_2|^{k_g} + |a_1|^{k_g} + |a_2|^{k_g})(|x_1 - x_2| + |a_1 - a_2|).$$

Thus, $v^0(x, a, \alpha)$ is locally Lipschitz in (x, a).

For each $N = 0, 1, 2, \ldots$, let

$$Z_N := \sup_{x_1, x_2, a_1, a_2} \frac{|v^N(x_1, a_1, \alpha) - v^N(x_2, a_2, \alpha)|}{(1 + |x_1|^{k_g} + |x_2|^{k_g} + |a_1|^{k_g} + |a_2|^{k_g})(|x_1 - x_2| + |a_1 - a_2|)}.$$

Then $Z_0 \leq C_0$.

For any $\alpha \in \mathcal{M}$ and $\beta \in \mathcal{M}$, let

$$\lambda_1(t) = -q_{\alpha\alpha}(a_1(t), u(t), w(t)), \quad \lambda_2(t) = -q_{\alpha\alpha}(a_2(t), u(t), w(t))$$

and $\quad p_\beta^1 = \dfrac{q_{\alpha\beta}(a_1(t), u(t), w(t))}{-q_{\alpha\alpha}(a_1(t), u(t), w(t))}, \quad p_\beta^2 = \dfrac{q_{\alpha\beta}(a_2(t), u(t), w(t))}{-q_{\alpha\alpha}(a_2(t), u(t), w(t))}.$

Using (3), it follows immediately that for all $t \geq 0$,

$$|\lambda_1(t) - \lambda_2(t)| \leq C|a_1 - a_2|$$

$$|p_\beta^1(t) - p_\beta^2(t)| \leq C|a_1 - a_2|.$$

(4)

Now for $N \geq 1$,

$$\left| J^N(x_1, a_1, \alpha, u(\cdot), w(\cdot), z(\cdot)) - J^N(x_2, a_2, \alpha, u(\cdot), w(\cdot), z(\cdot)) \right|$$

$$\leq \left| \int_0^\infty e^{-\rho t} \left(e^{-\int_0^t \lambda_1(r)dr} - e^{-\int_0^t \lambda_2(r)dr} \right) \left[G(x_1(t), a_1(t), u(t), w(t)) \right. \right.$$

$$\left. \left. + \lambda_1(t) \sum_{\beta \neq \alpha} v^{N-1}(x_1(t), a_1(t), \beta) p_\beta^1(t) \right] dt \right|$$

$$+ \int_0^\infty e^{-\rho t} \left(e^{-\int_0^t \lambda_2(r)dr} \right) |G(x_1(t), a_1(t), u(t), w(t))$$

$$- G(x_2(t), a_2(t), u(t), w(t))| dt$$

$$+ \int_0^\infty e^{-\rho t} \left(e^{-\int_0^t \lambda_2(r)dr} \right) |\lambda_1(t) - \lambda_2(t)| \sum_{\beta \neq \alpha} v^{N-1}(x_1(t), a_1(t), \beta) p_\beta^1(t) dt$$

$$+ \int_0^\infty e^{-\rho t} \left(e^{-\int_0^t \lambda_2(r)dr} \right) \lambda_2(t)$$

$$\left[\sum_{\beta \neq \alpha} \left[v^{N-1}(x_1(t), a_1(t), \beta) - v^{N-1}(x_2(t), a_2(t), \beta) \right] p_\beta^1(t) \right] dt$$

$$+ \int_0^\infty e^{-\rho t} \left(e^{-\int_0^t \lambda_2(r)dr} \right) \lambda_2(t) \sum_{\beta \neq \alpha} v^{N-1}(x_2(t), a_2(t), \beta) |p_\beta^1(t) - p_\beta^2(t)| dt$$

$$:= I_1 + I_2 + I_3 + I_4 + I_5.$$

In view of (4),

$$\left| e^{-\int_0^t \lambda_1(r)dr} - e^{-\int_0^t \lambda_2(r)dr} \right| \leq C \int_0^t |\lambda_1(r) - \lambda_2(r)| \, dr$$

$$\leq C \int_0^t |a_1 - a_2| dr = C|a_1 - a_2|t.$$

Thus, $I_1 \leq C_{I_1}(1 + |x_1|^{k_g} + |a_1|^{k_g})|a_1 - a_2|$ for a constant C_{I_1} independent of N.

By Assumption (A1), we can show that

$$I_2 \leq C_{I_2}(1 + |x_1|^{k_g} + |x_2|^{k_g} + |a_1|^{k_g} + |a_2|^{k_g})(|x_1 - x_2| + |a_1 - a_2|).$$

Similarly,

$$I_3 \leq C_{I_3}(1 + |x_1|^{k_g} + |a_1|^{k_g})|a_1 - a_2|$$

and

$$I_5 \leq C_{I_5}(1 + |x_2|^{k_g} + |a_2|^{k_g})|a_1 - a_2|.$$

We now estimate I_4. Note that

$$I_4 \leq Z_{N-1}(1 + |x_1|^{k_g} + |x_2|^{k_g} + |a_1|^{k_g} + |a_2|^{k_g}) \cdot$$
$$\cdot (|x_1 - x_2| + |a_1 - a_2|) \int_0^\infty e^{-\rho t} \left(e^{-\int_0^t \lambda_2(r)dr} \right) \lambda_2(t) dt.$$

Integration by parts leads to

$$\int_0^\infty e^{-\rho t} \left(e^{-\int_0^t \lambda_2(r)dr} \right) \lambda_2(t) dt = 1 - \rho \int_0^\infty e^{-\int_0^t (\lambda_2(r) + \rho r)dr} dt$$
$$\leq 1 - \rho \int_0^\infty e^{-\int_0^t (C_1 + \rho r)dr} dt \qquad (5)$$
$$= C_1/(C_1 + \rho) := \gamma_0 < 1,$$

where

$$C_1 := \sup_{\alpha, a, u, w} |q_{\alpha\alpha}(a, u, w)| < \infty.$$

Thus

$$I_4 \leq \gamma_0 Z_{N-1}(1 + |x_1|^{k_g} + |x_2|^{k_g} + |a_1|^{k_g} + |a_2|^{k_g})(|x_1 - x_2| + |a_1 - a_2|).$$

Combine the estimates on $I_1, \ldots, I_5$ to obtain

$$\left| J^N(x_1, a_1, \alpha, u(\cdot), w(\cdot), z(\cdot)) - J^N(x_2, a_2, \alpha, u(\cdot), w(\cdot), z(\cdot)) \right|$$
$$\leq (C_2 + \gamma_0 Z_{N-1})(1 + |x_1|^{k_g} + |x_2|^{k_g} + |a_1|^{k_g} + |a_2|^{k_g})$$
$$(|x_1 - x_2| + |a_1 - a_2|),$$

for some constant C_2. Thus, $v^N(x, a, \alpha)$ is locally Lipschitz. Moreover,

$$Z_0 \leq C_0 \text{ and } Z_N \leq C_2 + \gamma_0 Z_{N-1}, \text{ for } N \geq 1.$$

This implies that

$$Z_N \leq \gamma_0^N Z_0 + C_2 \sum_{i=0}^{N-1} \gamma_0^i \leq C_0 + C_2 \frac{1 - \gamma_0^N}{1 - \gamma_0} \leq C_0 + \frac{C_2}{1 - \gamma_0} := C_3.$$

If (b) holds, i.e., $\lim_{N \to \infty} v^N(x, a, \alpha) = v(x, a, \alpha)$, then $v(x, a, \alpha)$ is also locally Lipschitz. To prove (ii), it remains to verify (b). To this end, let $\mathcal{T}$

denote the following operator defined on the space of all continuous functions $V(x,a,\alpha)$:

$$\mathcal{T}V(x,a,\alpha)$$
$$= \inf_{(u(\cdot),w(\cdot))\in\mathcal{A}} \sup_{z(\cdot)\in\mathcal{Z}} \int_0^\infty e^{-\rho t}\left(e^{-\int_0^t \lambda(r)dr}\right)\bigg(G(x(t),a(t),u(t),w(t))$$
$$+\lambda(t)\sum_{\beta\neq\alpha}V(x(t),a(t),\beta)p_\beta(t)\bigg)dt.$$

Then by using (5), we can show

$$|\mathcal{T}V_1 - \mathcal{T}V_2| \leq \gamma_1|V_1 - V_2|,$$

with $0 < \gamma_1 < 1$. This means that $\mathcal{T}$ is a contraction operator. Moreover, $\mathcal{T}v^{N-1} = v^N$. By the fixed point theorem, we conclude that $v(x,a,\alpha) = \lim_{N\to\infty} v^N(x,a,\alpha)$.

Finally, it can be proved similarly as in [14] that the value function is a viscosity solution to the HJI equation. The uniqueness follows from [12], Theorem G.1. □

Theorem 2 (Verification Theorem). *Let $V(x,a,\alpha)$ denote a differentiable solution of the HJI equation (2) such that*

$$0 \leq V(x,a,\alpha) \leq C(1 + |x|^{k_g} + |a|^{k_g}).$$

Then:

(i) $V(x,a,\alpha) \leq \hat{J}(x,a,\alpha,u(\cdot),w(\cdot))$ *for all* $(u(\cdot),w(\cdot)) \in \mathcal{A}$;

(ii) *If* $(u^*(x,a,\alpha), w^*(x,a,\alpha))$ *is an admissible feedback control such that*

$$\min_{(u,w)\in\Gamma(\alpha)} \max_{z\in\Gamma_z} H(x,a,\alpha,v(x,a,\cdot),v_x(x,a,\alpha),v_a(x,a,\alpha),u,w,z)$$
$$= \max_{z\in\Gamma_z} H(x,a,\alpha,v(x,a,\cdot),v_x(x,a,\alpha),v_a(x,a,\alpha),u^*(x,a,\alpha),w^*(x,a,\alpha),z),\tag{6}$$

then

$$\hat{J}(x,a,\alpha,u^*(\cdot),w^*(\cdot)) = V(x,a,\alpha) = v(x,a,\alpha),$$

where
$$\Big(u^*(\cdot), w^*(\cdot)\Big) = \Big(u^*(x(\cdot), a(\cdot), \alpha(\cdot)), w^*(x(\cdot), a(\cdot), \alpha(\cdot))\Big).$$
Thus $(u^*(\cdot), w^*(\cdot))$ is optimal.

Proof. First of all, note that a differentiable solution to the HJI equation (2) is, in particular, a viscosity solution to that equation. The uniqueness of viscosity solutions in Theorem 1 implies

$$V(x, a, \alpha) = v(x, a, \alpha) \leq \hat{J}(x, a, \alpha, u(\cdot), w(\cdot)) \text{ for all } (u(\cdot), w(\cdot)) \in \mathcal{A}.$$

This proves (i).

To obtain (ii), it suffices to prove that

$$v(x, a, \alpha) = \sup_{z(\cdot) \in \mathcal{Z}} J(x, a, \alpha, u^*(\cdot), w^*(\cdot), z(\cdot)).$$

By virtue of the definition of value functions, that

$$v(x, a, \alpha) \leq \sup_{z(\cdot) \in \mathcal{Z}} J(x, a, \alpha, u^*(\cdot), w^*(\cdot), z(\cdot)).$$

It reminds to show

$$v(x, a, \alpha) \geq \sup_{z(\cdot) \in \mathcal{Z}} J(x, a, \alpha, u^*(\cdot), w^*(\cdot), z(\cdot)).$$

In view of (6), we can write the HJI equation (2) as follows:

$$\begin{aligned}\rho v(x, a, \alpha) = \max_{z \in \Gamma_z} \Big\{ & (u^*(x, a, \alpha) - z) v_x(x, a, \alpha) \\ & + m(\alpha, u^*(x, a, \alpha), w^*(x, a, \alpha)) v_a(x, a, \alpha) \\ & + G(x, a, u^*(x, a, \alpha), w^*(x, a, \alpha)) \\ & + Q(a, u^*(x, a, \alpha), w^*(x, a, \alpha)) v(x, a, \cdot)(\alpha) \Big\}.\end{aligned} \quad (7)$$

By Dynkin's formula, we have for each $z(\cdot) \in \mathcal{Z}$ and $T \geq 0$,

$$v(x,a,\alpha) = E\int_0^T e^{-\rho t}\Big\{\rho v(x(t),a(t),\alpha(t)) - (u^*(t) - z(t))v_x(x(t),a(t),\alpha(t))$$
$$+ m(\alpha(t),u^*(t),w^*(t))v_a(x(t),a(t),\alpha(t))$$
$$+ Q(a(t),u^*(t),w^*(t))v(x(t),a(t),\cdot)(\alpha(t))\Big\}dt$$
$$+ Ee^{-\rho T}v(x(T),a(T),\alpha(T))$$
$$\geq E\Big\{\int_0^T e^{-\rho t}G(x(t),a(t),u^*(t),w^*(t))dt + e^{-\rho T}v(x(T),a(T),\alpha(T))\Big\}.$$

The last inequality is due to (7). Observe that for any $T \geq 0$,

$$Ee^{-\rho T}v(x(T),a(T),\alpha(T)) \leq CEe^{-\rho T}(1 + |x(T)|^{k_g} + |a(T)|^{k_g}).$$

Note that $|x(T)| \leq C(1+|x|+T)$ and $|a(T)| \leq C(1+|a|+T)$ for some constant C. Hence

$$Ee^{-\rho T}v(x(T),a(T),\alpha(T)) \leq Ce^{-\rho T}(1 + T^{k_g} + |x|^{k_g} + |a|^{k_g}) \to 0,$$

as $T \to \infty$. Thus,

$$v(x,a,\alpha) \geq \lim_{T\to\infty} E\Big\{\int_0^T e^{-\rho t}G(x(t),a(t),u^*(t),w^*(t))dt$$
$$+ e^{-\rho T}v(x(T),a(T),\alpha(T))\Big\}$$
$$= J(x,a,\alpha,u^*(\cdot),w^*(\cdot),z(\cdot)),$$

for all $z(\cdot) \in \mathcal{Z}$. Thus

$$v(x,a,\alpha) \geq \sup_{z(\cdot)\in\mathcal{Z}} J(x,a,\alpha,u^*(\cdot),w^*(\cdot),z(\cdot)).$$

The proof of the theorem is completed. $\square$

4 A computational method

In the last section, we established a verification theorem (Theorem 2), which indicates that in order to solve for the optimal control problem, one only needs to solve the HJI equation. Very often, the closed form solutions of the corresponding HJI equations cannot be easily obtained. It is thus necessary and important to have efficient numerical algorithms available.

Inspired by the numerical methods introduced by Kushner in the 70's (see the more updated version of the treatment of Kushner and Dupuis [11] and the references therein), similar to the approach discussed in [2], we give a numerical scheme for solving the HJI equations in this section.

The basic idea behind this approach consists of using an approximation scheme for the partial derivatives of the value function $v(x, a, \alpha)$ within a finite grid of the state vector (x, a) and a finite grid for the control vector, which transforms the original optimization problem to an auxiliary discounted Markov decision process. This transformation allows us to apply the well-known techniques for solving the underlying optimization problems.

Let $\Delta x_j > 0$ and $\Delta a_i > 0$ denote the length of the finite difference interval of the variables x_j and a_i for $j = 1, \ldots, P$ and $i = 1, \ldots, M$.

Using a finite difference interval, we approximate the value function $v(x, a, \alpha)$ by a sequence of functions $v^\Delta(x, a, \alpha)$ and the partial derivatives $v_{x_j}(x, a, \alpha)$ by

$$\begin{cases} \dfrac{1}{\Delta x_j}(v^\Delta(x(\Delta x_j, +), a, \alpha) - v^\Delta(x, a, \alpha)) & \text{if } u_j - z_j \geq 0 \\ \dfrac{1}{\Delta x_j}(v^\Delta(x, a, \alpha) - v^\Delta(x(\Delta x_j, -), a, \alpha)) & \text{if } u_j - z_j < 0 \end{cases}$$

where

$$x(\Delta x_j, +) = (x_1, x_2, \ldots, x_{j-1}, x_j + \Delta x_j, x_{j+1}, \ldots, x_P)'$$

$$x(\Delta x_j, -) = (x_1, x_2, \ldots, x_{j-1}, x_j - \Delta x_j, x_{j+1}, \ldots, x_P)'.$$

Note that $m_i(\alpha, u, w) \geq 0$ for all i and all α, u, w. We approximate $v_{a_i}(x, a, \alpha)$ by

$$\dfrac{1}{\Delta a_i}(v^\Delta(x, a(\Delta a_i, +), \alpha) - v^\Delta(x, a, \alpha))$$

where

$$a(\Delta a_i, +) = (a_1, a_2, \ldots, a_{i-1}, a_i + \Delta a_i, a_{i+1}, \ldots, a_M)'.$$

Then, we have

$$(u_j - z_j)v_{x_j}(x, a, \alpha) \doteq \frac{|u_j - z_j|}{\Delta x_j} v^\Delta(x(\Delta x_j, +), a, \alpha)\chi_{\{u_j - z_j \geq 0\}}$$
$$+ \frac{|u_j - z_j|}{\Delta x_j} v^\Delta(x(\Delta x_j, -), a, \alpha)\chi_{\{u_j - z_j < 0\}}$$
$$- \frac{|u_j - z_j|}{\Delta x_j} v^\Delta(x, a, \alpha),$$

where χ_F is the indicator function of a set F. Similarly,

$$m_i(\alpha, u, w)v_{a_i}(x, a, \alpha)$$
$$\doteq \frac{m_i(\alpha, u, w)}{\Delta a_i} v^\Delta(x, a(\Delta a_i, +), \alpha) - \frac{m_i(\alpha, u, w)}{\Delta a_i} v^\Delta(x, a, \alpha).$$

With these approximations, we can "rewrite" the HJI equation (2) in terms of $v^\Delta(x, a, \alpha)$ as:

$$v^\Delta(x, a, \alpha) = \min_{(u,w)\in\Gamma(\alpha)} \max_{z\in\Gamma_z} \left[\rho + |q_{\alpha\alpha}(a, u, w)|\right.$$
$$+ \sum_{j=1}^{P} \frac{|u_j - z_j|}{\Delta x_j} + \sum_{i=1}^{M} \frac{m_i(\alpha, u, w)}{\Delta a_i}\bigg]^{-1}$$
$$\times \left\{\sum_{j=1}^{P} \frac{|u_j - z_j|}{\Delta x_j}\left(v^\Delta(x(\Delta x_j, +), a, \alpha)\chi_{\{u_j - z_j \geq 0\}}\right.\right.$$
$$+ v^\Delta(x(\Delta x_j, -), a, \alpha)\chi_{\{u_j - z_j < 0\}}\bigg)$$
$$+ \sum_{i=1}^{M} \frac{m_i(\alpha, u, w)}{\Delta a_i} v^\Delta(x, a(\Delta a_i, +), \alpha) + G(x, a, u, w)$$
$$+ \sum_{\beta \neq \alpha} q_{\alpha\beta}(a, u, w)v^\Delta(x, a, \beta)\bigg\}.$$

(8)

Using the same argument as that of [2, 11], we obtain the following procedure:

$$v^\Delta(x, a, \alpha) = \min_{(u,w)\in\Gamma(\alpha)} \max_{z\in\Gamma_z} \left\{\frac{G(x, a, u, w)}{\Psi(x, a, u, w)\left(1 + \frac{\rho}{\Psi(x,a,u,w)}\right)} + \frac{1}{\left(1 + \frac{\rho}{\Psi(x,a,u,w)}\right)}\right.$$
$$\times \bigg[\sum_{j=1}^{P} P^\alpha_{x,a;x_j,a}\left(v^\Delta(x(\Delta x_j, +), a, \alpha)\chi_{\{u_j - z_j \geq 0\}}\right.$$
$$+ v^\Delta(x(\Delta x_j, -), a, \alpha)\chi_{\{u_j - z_j < 0\}}\bigg)$$
$$+ \sum_{i=1}^{M} P^\alpha_{x,a;x,a_i} v^\Delta(x, a(\Delta a_i, +), \alpha) + \sum_{\beta \neq \alpha} P^\alpha_{\alpha;\beta} v^\Delta(x, a, \beta)\bigg]\bigg\},$$

(9)

where

$$\Psi(x,a,u,w) = \left(|q_{\alpha\alpha}(a,u,w)| + \sum_{j=1}^{P} \frac{|u_j - z_j|}{\Delta x_j} + \sum_{i=1}^{M} \frac{m_i(\alpha,u,w)}{\Delta a_i}\right)$$

$$P^{\alpha}_{x,a;x_j,a} = \frac{|u_j - z_j|}{\Delta x_j}(\Psi(x,a,u,w))^{-1}$$

$$P^{\alpha}_{x,a;x,a_i} = \frac{m_i(\alpha,u,w)}{\Delta a_i}(\Psi(x,a,u,w))^{-1}$$

$$P^{\alpha}_{\alpha;\beta} = \frac{q_{\alpha\beta}(a,u,w)}{\Psi(x,a,u,w)}.$$

Note that

$$P^{\alpha}_{x,a;x_j,a} \geq 0, \ P^{\alpha}_{x,a;x,a_i} \geq 0, \ P^{\alpha}_{\alpha;\beta} \geq 0$$

and

$$\sum_{j=1}^{P} P^{\alpha}_{x,a;x_j,a} + \sum_{i=1}^{M} P^{\alpha}_{x,a;x,a_i} + \sum_{\beta \neq \alpha} P^{\alpha}_{\alpha;\beta} = 1.$$

Therefore equation (9) can be interpreted as the HJI equation for a discounted Markovian decision problem.

The solution of this discounted Markov decision process may be obtained by either successive approximation or policy improvement techniques; see [2, 11] for related discussions.

In the next theorem, we show that $v^{\Delta}(x,a,\alpha)$ converges to $v(x,a,\alpha)$ as the step size Δx_j and Δa_i go to zero. For simplicity, we only consider the case that $\Delta x_j = \Delta a_i = \Delta$ for $j = 1,2,\ldots,P$ and $i = 1,2,\ldots,M$.

Theorem 3 (Convergence of the approximation scheme). *Assume there exists a constant C such that*

$$0 \leq v^{\Delta}(x,a,\alpha) \leq C(1 + |x|^{k_g} + |a|^{k_g}).$$

Then,

$$\lim_{\Delta \to 0} v^{\Delta}(x,a,\alpha) = v(x,a,\alpha). \tag{10}$$

Proof. First of all, note that the equation (8) can be written as

$$v^{\Delta}(x,a,\alpha) = \tilde{T}v^{\Delta}(x,a,\alpha), \tag{11}$$

for an operator $\tilde{\mathcal{T}}$. It is not difficult to check that for each $\Delta > 0$, the operator $\tilde{\mathcal{T}}$ is a contraction mapping. This, in turn, implies that the equation (8) has only one solution $v^\Delta(x, a, \alpha)$. Moreover, by the iterative scheme,

$$v_0^\Delta(x, a, \alpha) = 0, \text{ and } v_{N+1}^\Delta(x, a, \alpha) := \tilde{\mathcal{T}} v_N^\Delta(x, a, \alpha), \ N = 1, 2, \ldots,$$

we can show that the solution to (11) is continuous.

Note that for any fixed $\Delta > 0$,

$$0 < \rho \leq \rho + |q_{\alpha\alpha}(a, u, w)| + \sum_{j=1}^{P} \frac{|u_j - z_j|}{\Delta x_j} + \sum_{i=1}^{M} \frac{m_i(\alpha, u, w)}{\Delta a_i} \leq C(1 + \Delta^{-1}).$$

It yields that the following equation is equivalent to (8). That is, a solution to any one of them is a solution to the other:

$$\begin{aligned}
0 = \min_{(u,w)\in\Gamma(\alpha)} \max_{z \in \Gamma_z} &\bigg\{ \sum_{j=1}^{P} \frac{|u_j - z_j|}{\Delta x_j} \Big([v^\Delta(x(\Delta x_j, +), a, \alpha) - v^\Delta(x, a, \alpha)] \\
&\times \chi_{\{u_j - z_j \geq 0\}} + [v^\Delta(x(\Delta x_j, -), a, \alpha) - v^\Delta(x, a, \alpha)] \chi_{\{u_j - z_j < 0\}} \Big) \\
&+ \sum_{i=1}^{M} \frac{m_i(\alpha, u, w)}{\Delta a_i} \Big(v^\Delta(x, a(\Delta a_i, +), \alpha) - v^\Delta(x, a, \alpha) \Big) \\
&+ G(x, a, u, w) + \sum_{\beta \neq \alpha} q_{\alpha\beta}(a, u, w) \\
&\times [v^\Delta(x, a, \beta) - v^\Delta(x, a, \alpha)] - \rho v^\Delta(x, a, \alpha) \bigg\}.
\end{aligned} \tag{12}$$

Let

$$v^*(x, a, \alpha) := \limsup_{\delta \to 0} \left[\limsup_{\Delta \to 0} \left[\sup\{v^\Delta(x', a', \alpha) : |(x, a) - (x', a')| \leq \delta\} \right] \right]$$

and

$$v_*(x, a, \alpha) := \liminf_{\delta \to 0} \left[\liminf_{\Delta \to 0} \left[\inf\{v^\Delta(x', a', \alpha) : |(x, a) - (x', a')| \leq \delta\} \right] \right].$$

Then it follows that $v^*(x, a, \alpha)$ is upper semicontinuous, $v_*(x, a, \alpha)$ is lower semicontinuous, and $v^*(x, a, \alpha) \geq v_*(x, a, \alpha)$.

To obtain the convergence result, it remains to derive the reverse inequality, $v^*(x, a, \alpha) \leq v_*(x, a, \alpha)$. In fact, we need only show that $v^*(x, a, \alpha)$ and $v_*(x, a, \alpha)$ are viscosity subsolution and viscosity supersolution to (2), respectively. This can be done as in [11], Theorem 14.3.1. Subsequently, by virtue of [12], Theorem G.1, we have $v^*(x,a,\alpha) \leq v_*(x,a,\alpha)$. Hence, $v^*(x,a,\alpha) = v_*(x,a,\alpha) = v(x,a,\alpha)$ as desired. □

Remark: An alternative approach for proving Theorem 3 is to apply the method of weak convergence via a Markov chain approximation in the setup of [11]. However, special care must be taken to treat the minimax robust control problem.

5 Conclusions

We developed a robust production planning model for manufacturing systems under unknown and un-modeled disturbance (demand) by using minimax cost functions. The model discussed here is a slight modification of that of [7]. Due to the page limitation, the fully detailed proofs are not given in [7], whereas they are verbatim spelled out in the current work. We showed that the value function is the unique viscosity solution of the associated HJI equations, derived a verification theorem that provides a sufficient condition for optimal control policies, and developed a numerical method for solving the HJI equations. It was shown that the solution of the numerical procedure converges to the value function as the step size goes to zero.

We only treated relatively simple model. It appears that it is possible to generalize the results to more complex models such as flowshops or even jobshops (see [12] among others).

Finally, we notice that the numerical scheme given in this paper can be used effectively for medium sized problems. For large system problems with large dimensional state space, one has to resort to other approximation methods, for example, the hierarchical decomposition approach. The idea of this approach is to reduce the large complex problem into manageable approximate problems or subproblems, to solve these sub-problems, and to construct a solution of the original problem from the solutions of these simpler problems. See [12] for more details on this approach.

References

[1] Başar, T. and Bernhard, P., H^∞ - *Optimal Control and Related Minimax Design Problems*, Birkhäuser, Boston, 1991.

[2] Boukas, E. K., Techniques for flow control and preventive maintenance in manufacturing systems, *Control and Dynamic Systems*, Vol. 48, pp. 327-366, 1991.

[3] Boukas, E. K., Control of systems with controlled jump Markov disturbances, to appear in *International Journal of Control Theory and Advanced Technology*, 1993.

[4] Boukas, E. K., A stochastic modeling for flexible manufacturing systems, submitted to *IEEE Transaction Robotics and Automation*.

[5] Boukas, E. K. and Haurie, A., Manufacturing flow control and preventive maintenance: A stochastic control approach, *IEEE Transactions on Automatic Control*, Vol. 35, No. 9, pp. 1024-1031, 1990.

[6] Boukas, E. K., Zhu, Q., and Zhang, Q., A piecewise deterministic Markov process model for flexible manufacturing systems with preventive maintenance, *Journal of Optimization Theory and Applications*, 1994. Vol. 81, No. 2, pp. 259-275, 1994.

[7] Boukas, E.K., Zhang, Q. and Yin, G., Robust production and maintenance planning in stochastic manufacturing systems, *IEEE Trans. Automat. Control* 40 (1995), 1098-1102.

[8] Davis, M.H.A., *Markov Models and Optimization*, Chapman & Hall, New York, NY, 1993.

[9] Fleming, W.H. and Soner, H.M., *Controlled Markov Processes and Viscosity Solutions*, Springer-Verlag, New York, 1992.

[10] Hu, J. Q., Vakili, P. and Yu, G.X., Optimality of hedging point policies in the production control of failure prone manufacturing systems, *IEEE Transactions on Automatic Control*, Vol. 39, pp. 1875-1880, 1994.

[11] Kushner, H.J. and Dupuis, P.G. *Numerical Methods for Stochastic Control Problems in Continuous Time*, Springer-Verlag, New York, 1992.

[12] Sethi, S.P. and Zhang, Q., *Hierarchical Decision Making in Stochastic Manufacturing Systems*, Birkhäuser, Boston, 1994.

[13] Soner, H.M., Singular perturbations in manufacturing, *SIAM J. Control and Optimization*, Vol. 31, pp. 132-146, 1993.

[14] Zhang, Q., Risk sensitive production planning of stochastic manufacturing systems: A singular perturbation approach, *SIAM J. Control and Optimization*, Vol. 33, No. 2, pp. 498-527, 1995.

E.K. BOUKAS
MECHANICAL ENGINEERING DEPARTMENT
ÉCOLE POLYTECHNIQUE DE MONTRÉAL AND THE GERAD
MONTRÉAL, QUE., CANADA

Q. ZHANG
DEPARTMENT OF MATHEMATICS
UNIVERSITY OF GEORGIA
ATHENS, GA 30602

G. YIN
DEPARTMENT OF MATHEMATICS
WAYNE STATE UNIVERSITY
DETROIT, MI 48202

Lecture Notes in Control and Information Sciences

Edited by M. Thoma

1992–1996 Published Titles:

Vol. 167: Rao, Ming
Integrated System for Intelligent Control
133 pp. 1992 [3-540-54913-7]

Vol. 168: Dorato, Peter; Fortuna, Luigi; Muscato, Giovanni
Robust Control for Unstructured Perturbations: An Introduction
118 pp. 1992 [3-540-54920-X]

Vol. 169: Kuntzevich, Vsevolod M.; Lychak, Michael
Guaranteed Estimates, Adaptation and Robustness in Control Systems
209 pp. 1992 [3-540-54925-0]

Vol. 170: Skowronski, Janislaw M.; Flashner, Henryk; Guttalu, Ramesh S. (Eds)
Mechanics and Control. Proceedings of the 4th Workshop on Control Mechanics, January 21-23, 1991, University of Southern California, USA
302 pp. 1992 [3-540-54954-4]

Vol. 171: Stefanidis, P.; Paplinski, A.P.; Gibbard, M.J.
Numerical Operations with Polynomial Matrices: Application to Multi-Variable Dynamic Compensator Design
206 pp. 1992 [3-540-54992-7]

Vol. 172: Tolle, H.; Ersü, E.
Neurocontrol: Learning Control Systems Inspired by Neuronal Architectures and Human Problem Solving Strategies
220 pp. 1992 [3-540-55057-7]

Vol. 173: Krabs, W.
On Moment Theory and Controllability of Non-Dimensional Vibrating Systems and Heating Processes
174 pp. 1992 [3-540-55102-6]

Vol. 174: Beulens, A.J. (Ed.)
Optimization-Based Computer-Aided Modelling and Design. Proceedings of the First Working Conference of the New IFIP TC 7.6 Working Group, The Hague, The Netherlands, 1991
268 pp. 1992 [3-540-55135-2]

Vol. 175: Rogers, E.T.A.; Owens, D.H.
Stability Analysis for Linear Repetitive Processes
197 pp. 1992 [3-540-55264-2]

Vol. 176: Rozovskii, B.L.; Sowers, R.B. (Eds)
Stochastic Partial Differential Equations and their Applications. Proceedings of IFIP WG 7.1 International Conference, June 6-8, 1991, University of North Carolina at Charlotte, USA
251 pp. 1992 [3-540-55292-8]

Vol. 177: Karatzas, I.; Ocone, D. (Eds)
Applied Stochastic Analysis. Proceedings of a US-French Workshop, Rutgers University, New Brunswick, N.J., April 29-May 2, 1991
317 pp. 1992 [3-540-55296-0]

Vol. 178: Zolésio, J.P. (Ed.)
Boundary Control and Boundary Variation. Proceedings of IFIP WG 7.2 Conference, Sophia-Antipolis, France, October 15-17, 1990
392 pp. 1992 [3-540-55351-7]

Vol. 179: Jiang, Z.H.; Schaufelberger, W.
Block Pulse Functions and Their Applications in Control Systems
237 pp. 1992 [3-540-55369-X]

Vol. 180: Kall, P. (Ed.)
System Modelling and Optimization.
Proceedings of the 15th IFIP Conference,
Zurich, Switzerland, September 2-6, 1991
969 pp. 1992 [3-540-55577-3]

Vol. 181: Drane, C.R.
Positioning Systems - A Unified Approach
168 pp. 1992 [3-540-55850-0]

Vol. 182: Hagenauer, J. (Ed.)
Advanced Methods for Satellite and Deep
Space Communications. Proceedings of
an International Seminar Organized by
Deutsche Forschungsanstalt für Luft-und
Raumfahrt (DLR), Bonn, Germany,
September 1992
196 pp. 1992 [3-540-55851-9]

Vol. 183: Hosoe, S. (Ed.)
Robust Control. Proceesings of a Workshop
held in Tokyo, Japan, June 23-24, 1991
225 pp. 1992 [3-540-55961-2]

Vol. 184: Duncan, T.E.; Pasik-Duncan, B.
(Eds)
Stochastic Theory and Adaptive Control.
Proceedings of a Workshop held in
Lawrence, Kansas, September 26-28,
1991
500 pp. 1992 [3-540-55962-0]

Vol. 185: Curtain, R.F. (Ed.); Bensoussan,
A.; Lions, J.L.(Honorary Eds)
Analysis and Optimization of Systems:
State and Frequency Domain Approaches
for Infinite-Dimensional Systems.
Proceedings of the 10th International
Conference, Sophia-Antipolis, France, June
9-12, 1992.
648 pp. 1993 [3-540-56155-2]

Vol. 186: Sreenath, N.
Systems Representation of Global Climate
Change Models. Foundation for a Systems
Science Approach.
288 pp. 1993 [3-540-19824-5]

Vol. 187: Morecki, A.; Bianchi, G.;
Jaworeck, K. (Eds)
RoManSy 9: Proceedings of the Ninth
CISM-IFToMM Symposium on Theory and
Practice of Robots and Manipulators.
476 pp. 1993 [3-540-19834-2]

Vol. 188: Naidu, D. Subbaram
Aeroassisted Orbital Transfer: Guidance
and Control Strategies
192 pp. 1993 [3-540-19819-9]

Vol. 189: Ilchmann, A.
Non-Identifier-Based High-Gain Adaptive
Control
220 pp. 1993 [3-540-19845-8]

Vol. 190: Chatila, R.; Hirzinger, G. (Eds)
Experimental Robotics II: The 2nd
International Symposium, Toulouse,
France, June 25-27 1991
580 pp. 1993 [3-540-19851-2]

Vol. 191: Blondel, V.
Simultaneous Stabilization of Linear
Systems
212 pp. 1993 [3-540-19862-8]

Vol. 192: Smith, R.S.; Dahleh, M. (Eds)
The Modeling of Uncertainty in Control
Systems
412 pp. 1993 [3-540-19870-9]

Vol. 193: Zinober, A.S.I. (Ed.)
Variable Structure and Lyapunov Control
428 pp. 1993 [3-540-19869-5]

Vol. 194: Cao, Xi-Ren
Realization Probabilities: The Dynamics of
Queuing Systems
336 pp. 1993 [3-540-19872-5]

Vol. 195: Liu, D.; Michel, A.N.
Dynamical Systems with Saturation
Nonlinearities: Analysis and Design
212 pp. 1994 [3-540-19888-1]

Vol. 196: Battilotti, S.
Noninteracting Control with Stability for
Nonlinear Systems
196 pp. 1994 [3-540-19891-1]

Vol. 197: Henry, J.; Yvon, J.P. (Eds)
System Modelling and Optimization
975 pp approx. 1994 [3-540-19893-8]

Vol. 198: Winter, H.; Nüßer, H.-G. (Eds)
Advanced Technologies for Air Traffic Flow Management
225 pp approx. 1994 [3-540-19895-4]

Vol. 199: Cohen, G.; Quadrat, J.-P. (Eds)
11th International Conference on
Analysis and Optimization of Systems –
Discrete Event Systems: Sophia-Antipolis,
June 15–16–17, 1994
648 pp. 1994 [3-540-19896-2]

Vol. 200: Yoshikawa, T.; Miyazaki, F. (Eds)
Experimental Robotics III: The 3rd International Symposium, Kyoto, Japan, October 28-30, 1993
624 pp. 1994 [3-540-19905-5]

Vol. 201: Kogan, J.
Robust Stability and Convexity
192 pp. 1994 [3-540-19919-5]

Vol. 202: Francis, B.A.; Tannenbaum, A.R. (Eds)
Feedback Control, Nonlinear Systems, and Complexity
288 pp. 1995 [3-540-19943-8]

Vol. 203: Popkov, Y.S.
Macrosystems Theory and its Applications: Equilibrium Models
344 pp. 1995 [3-540-19955-1]

Vol. 204: Takahashi, S.; Takahara, Y.
Logical Approach to Systems Theory
192 pp. 1995 [3-540-19956-X]

Vol. 205: Kotta, U.
Inversion Method in the Discrete-time Nonlinear Control Systems Synthesis Problems
168 pp. 1995 [3-540-19966-7]

Vol. 206: Aganovic, Z.;.Gajic, Z.
Linear Optimal Control of Bilinear Systems with Applications to Singular Perturbations and Weak Coupling
133 pp. 1995 [3-540-19976-4]

Vol. 207: Gabasov, R.; Kirillova, F.M.; Prischepova, S.V.
Optimal Feedback Control
224 pp. 1995 [3-540-19991-8]

Vol. 208: Khalil, H.K.; Chow, J.H.; Ioannou, P.A. (Eds)
Proceedings of Workshop on Advances in Control and its Applications
300 pp. 1995 [3-540-19993-4]

Vol. 209: Foias, C.; Özbay, H.; Tannenbaum, A.
Robust Control of Infinite Dimensional Systems: Frequency Domain Methods
230 pp. 1995 [3-540-19994-2]

Vol. 210: De Wilde, P.
Neural Network Models: An Analysis
164 pp. 1996 [3-540-19995-0]

Vol. 211: Gawronski, W.
Balanced Control of Flexible Structures
280 pp. 1996 [3-540-76017-2]

Vol. 212: Sanchez, A.
Formal Specification and Synthesis of Procedural Controllers for Process Systems
248 pp. 1996 [3-540-76021-0]

Vol. 213: Patra, A.; Rao, G.P.
General Hybrid Orthogonal Functions and their Applications in Systems and Control
144 pp. 1996 [3-540-76039-3]